Implementing and Improving ITSM

A leader's guide to implementing an Enterprise IT Service Management

Jeffrey Tefertiller
Service Management Leadership

Implementing and Improving ITSM
Copyright © 2018 by Jeffrey Tefertiller
Service Management Leadership

ITIL© is a registered trademark of Axelos Limited. All rights reserved.

Quoted ITIL text distinguished in *italic font* is from the Axelos ITIL© books. Please refer to the ITIL© books for further explanation.

Implementing and Improving ITSM

Table of Contents

Introduction 4

Section One

Chapter 1 – The Framework 9

    1.1   Framework Scope 12

    1.2   Alignment with Organization Strategy 13

Chapter 2 – Establishing a SMO 20

Chapter 3 – Process Control 33

Chapter 4 – Roles and Responsibilities 37

Chapter 5 – Process Design Steps 50

Section Two

Chapter 6 – Service Strategy 54

Chapter 7 – Service Design 79

Chapter 8 – Service Transition 128

Chapter 9 – Service Operations 172

Chapter 10 – Continual Service Improvement 220

Section Three

Chapter 11 – Quality, Performance & Metrics 228

    11.1 Service Quality Management 229

    11.2 Service Performance 241

    11.3 Assess, Measurements, and Metrics 246

Chapter 12 – Organizational Considerations 248

Chapter 13 – Risk Management 250

Glossary of Terms 251

Introduction

Providing service excellence is a worthy goal for organizations because a proactive Service environment is less expensive to run than a reactive Service environment. We see this principle at work with our automobiles. A proactive repair (preventive maintenance or even new tires) offers a better cost point, higher quality because you have options, and less risk than if you have a reactive repair on the side of the road where options are limited. This exact scenario plays out regularly in Information Technology (IT) organizations. Service excellence is the proactive goal for most organizations.

This book will provide guidance to implement standard procedures for managing IT Services and provide an example of the integration of Best Practices, frameworks, and standards that define an organization-wide Service Management approach. This book offers a 'Service-oriented' Framework that focuses on creating and managing Services throughout the Service Lifecycle. It will align and integrate processes for Service Management and define processes at a high level, describing more of the *what* than the *how*. This approach enables cross-functional teams the ability to create and improve processes in the common pursuit of Service excellence.

As is the case with all our writings, this Framework is targeted toward leaders to help initiate, lead, or improve an ITSM Program. Few organizations have implemented all the processes. No matter how mature your organization's ITSM Program, this book is for leaders seeking solutions. This is a leadership guide that will aid leaders responsible for administering their firms' ITSM Programs.

How to use this IT Service Management Framework (ITSM framework):

There is a lack of ITSM-specific standard terminology for much of this content within most organizations, both in the private and public sectors. Therefore, it is necessary to define a few key terms used in this book.

- **Service** is a means of delivering value using people, processes, and technology perceived by Customers and Users as a self-contained, single, coherent entity that enables them to achieve business objectives.

- **Policies** help with governance and are formally documented expectations and intentions of the organization's leadership. Policies are used to direct decisions, ensure consistent and appropriate development, and direct the implementation of processes, standards, roles, activities, IT infrastructure, and other areas needing governance.
- **Process** is a structured set of activities designed to accomplish a specific objective. A process is made up of interconnecting activities that use inputs, controls, and enablers (typically tools) to produce a defined output (or set of outputs). A process may draw from any of the roles, responsibilities, tools and leadership controls required to reliably deliver the outputs and comprises specific procedures to accomplish this activity.
- **Procedure** is a document containing detailed steps that specify how to perform process steps. Procedures are defined as a part of processes and are less granular in nature than work instructions.

This Framework may be used to address Services and processes owned and managed by your organization and is applicable to internal Service Providers, external Service Providers, and shared Service Providers. The Framework includes:

1. Guidance on IT Service alignment with business customers
2. Processes – Enterprise-wide processes defined at a high level and guidance to establish authoritative Service Owners, Process Owners, and Process Managers, etc.
3. Purpose and Scope – The purpose and scope of each process in the lifecycle
4. Process Workflow Guidance - Mapped activities and supporting explanation
5. Metrics – Recommendations on the use of metrics as actionable items
6. Roles and Responsibilities – Defined responsibilities of related ITSM roles
7. Service Quality Management Approach – Describes the approach to establish, implement, and maintain Service quality
8. IT Performance Management Guidance – Describes an approach consisting of activities that focus on up-front planning and aligning IT with defined goals
9. Process Capability Assessment Information – Defines an approach to evaluate and measure the competency of a process to meet its intended purpose and outcomes

Implementing and Improving ITSM

This Framework is not a detailed implementation plan. While this book contains steps for process design at a high, leadership level, it is not meant as a comprehensive project plan or as an overall ITSM implementation plan. This Framework describes the "Who," "What," and "How," not the exhaustive manner of implementation.

> Since the goal of this book is to help fellow leaders understand and implement better ITSM programs, keep an eye out for blue text boxes like this one. These will be topics and concepts applying ITSM principles for leaders.

This book is divided into easy-to-follow sections. Section One discusses the enablers needed to develop your organization's Framework. Section Two is an end-to-end process list – including Purpose, Scope, Expected Outcomes and Benefits, Process Workflow, and Process Activities – for all the key processes in the Service Lifecycle. Section Three includes areas that need to be considered if your Framework is to be successful.

Section One

Chapter 1 – The Framework

There are many IT frameworks available throughout the public and private sectors. However, this ITSM Framework provides an integrated framework to provide guidance and establish the structure, documentation, and roles and responsibilities to plan, implement, monitor and improve an ITSM Program.

The purpose of this Framework is to provide guidance for planning, implementing, monitoring, and improving Service Management initiatives for all IT Services across your organization. This Framework establishes a documented and clear foundation for enterprise process implementation and execution across the organization. This Framework will discuss the use of ITSM methods that include specific ways to optimize the planning and implementation of ITIL© best practices in your organization. This ITSM Framework addresses the ongoing alignment of the delivery of IT Services closely with the needs of the business customers. In addition, the Framework may help establish enterprise-wide processes allowing the effective use of ITSM technology.

This Framework will introduce the ITSM processes, including the relationship between processes, enabling IT Services to effectively support the business customers. We will begin the discussion with the orange section below (Section One – ITSM Framework Enablers).

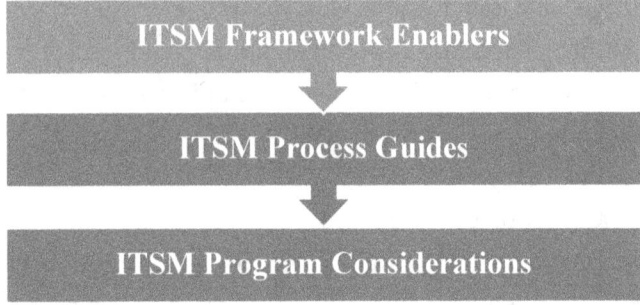

ITSM process and service improvement initiatives should align with the ITSM Framework. To that end, this book will:

Implementing and Improving ITSM

- Define the best practices that drive the implementation of the Framework
- Define the overall structure of the Framework, including the ITSM Service Lifecycle phases and processes
- Provide a general overview of processes in terms of purpose, scope, benefits, terminology, and roles and responsibilities
- Define the controls Framework required to meet compliance with the agreed standards needed for most organizations in all industries
- Define the recommended interfaces between the Service Lifecycle phases and processes
- Recommend milestones for process implementation and service improvements

The goal of the Framework is to provide guidance for a successful alignment of the delivery of IT Services with your organization's vision and strategy. A successful ITSM Program integrates people, processes, and technology in a concerted effort to deliver value to business customers.

1.1 Framework Scope

The scope of the Framework applies to all IT services and the ITSM processes that support those services. Service and process initiatives should align the Framework to the goals and objectives of the business customers.

As your Framework matures, the goal is to develop, publish, and promote an organizationally accepted process architecture and Service Management Program for strategic, tactical, and operational use by your organization to enable better decision-making and efficient Service delivery.

1.2 Alignment with Organization Strategy

To ensure that the Framework contains the strategic elements necessary to provide a focused and purposeful Service Management Framework, it must be continually aligned with key organizational strategic plans and initiatives.

This Framework provides a structure to enable the delivery and support of Services to accomplish the vision and strategy objectives of your organization.

Framework benefits for organizations:
- Provides a single, definable, repeatable, and scalable documented Framework
- Clearly identifies roles and responsibilities
- Adoption enables your organization to provide higher Service-quality and availability levels, improve alignment between Service Provider and business customers, and better ensure security and capability of the information enterprise (Warranty processes to be designed into Services)
- Enables better decision-making at all levels of the organization by identifying relationships between all phases and processes throughout the Service Lifecycle
- Services will be measured transparently to all stakeholders

Framework benefits for organizations:

- The costs associated with the entire Service Lifecycle are understood
- Compliance will use repeatable audit processes
- Better understanding of IT Services and the value derived from each Service, both from the provider as well as the business customer perspective
- Supports ability of IT to measure and improve Service performance
- Improved business customer satisfaction through an efficient approach to Service delivery and support
- Secure information and data exchange
- Enhanced ability to mature and increase performance based on feedback incorporated into the processes and services
- Cost effectiveness and efficiency are realized by identifying duplications for upgrade or removal
- Recommended milestones for process implementation and service improvements
- Improved business customer satisfaction through an efficient approach to Service delivery and support
- Secure information and data exchange
- Enhanced ability to mature and increase performance based on feedback incorporated into the processes and services
- Cost effectiveness and efficiency are realized by identifying duplications for upgrade or removal
- Recommended milestones for process implementation and service improvements

There are several key Critical Success Factors (CSFs) for Framework adoption:
- Form a Service Management Office (SMO) – Create an office to provide guidance on enhancing and maintaining the Framework
- Create strategic vision – Clearly identify the gap that this ITSM initiative is trying to close
- Communicate the vision – Create a proactive and directed training and communications plan for the adoption of the Framework
- Identify short-term wins – Identify and execute to improve on issues related to how services are currently provided
- Establish Key Performance Indicators (KPIs) – KPIs are required to measure each CSF
- Align KPIs – KPIs are required to align the IT organization's goals and plans with strategic objectives of the business customers

General Framework Guidance:
1. **Each process should have a single Process Owner, accountable for process quality and integrity** - Multiple ownership is divided ownership, and it creates a less-optimized process and increases the likelihood of overlapping responsibilities or gaps, and areas of the process being measured.
2. Processes and Services should be designed with sufficient flexibility to ensure not only that the current needs of the organization are addressed, but that future requirements of technology and capacity are anticipated and accounted for as part of the Service Lifecycle. - Organizational requirements change rapidly in the current technology environment and current processes must support nimble Service development and implementation.
3. **Even in the most distributed organization, there should be consistent processes.** - To provide the appropriate levels of governance and measure the overall effect of implementing the Framework, consistent processes are necessary.

Below are the **SIX** high-level steps (not intended to be exhaustive) for how to implement a Service or process improvement effort. This is a leadership viewpoint, so it is not granular.

These steps correlate to the ITSM Service Lifecycle phases: Service Strategy, Service Design, Service Transition, Service Operations, and Continual Service Improvement.

1. Define Outcomes for Implementation

Define CSFs and KPIs for each process
Create the process-implementation roadmap, including order of implementation.

2. Develop a flexible governance structure and align with the Service Lifecycle

Define the Roles and Responsibilities for each Service Lifecycle phase
Define the Governance workflow
Develop the Governance Organizational Change Management (Communications) Plan
Develop the ITSM process architectures for all stages of implementation (pilot, early life support, and full operational support) required and align to your organization

3. Define organizational process ownership

Align areas best suited for ownership of processes to the organization
Identify Process Owners
Create the implementation roadmap to ensure Process Owners are available to give guidance to other Process Owners at appropriate phases of implementation

4. Develop a training plan

Develop framework training requirements
Identify required training levels according to Service Lifecycle role:
Domain Owner
Process Owner
Process Manager
Service Owner
Service Manager
General Staff
Other Stakeholders as necessary

5. Determine Risk Strategy

Define criteria for decision-making
Identify risk assessment processes
Define and implement a risk register
Identify risk mitigation activities

6. Define collaborative relationships

Service Management Office (SMO)
Project Management Office (PMO)
Application Development Organization(s)
Maintenance and Operations Organization(s)
Other organizations as necessary

Chapter 2 – Establishing a Service Management Office (SMO)

Within most organizations, the IT Service Management (ITSM) activities usually begin as a set of individual efforts to manage and improve IT Service delivery and support. Most organizations fall into this category, especially those that are large and organizationally dispersed with subordinate Services. Issues arise when competing interests are not holistically aligned within the organization.

Best Practice guides organizations to position the SMO with authority and responsibility to execute and implement the ITSM Program, with far greater scope than just employing the SMO to be a consulting office. **The SMO needs to support ITSM adoption and champion Continual Service Improvement (CSI).** The SMO should embody the governing body and add value in an advisory and support capacity.

> Every organization should have a single authority from which to obtain IT Service Management training, artifacts, templates etc.

Implementing and Improving ITSM

The SMO serves as a catalyst, facilitating ownership of process, technology, and Services, informing and training the organization on ITSM practices and procedures, and ultimately driving improvements within the ITSM Program.

> **The end state of all ITSM initiatives should be effective, efficient, and secure IT Services.**

The SMO:	
	Champions end-to-end standards for the entire IT organization
	Collects, assembles, and distributes Service Management training and knowledge throughout the IT organization
	Monitors and improves processes and Services using defined metrics and a standardized, consistent approach
	Drives shared ITSM capabilities (e.g. monitoring tools, CMDB, ITSM tools)

Implementing and Improving ITSM

> The goal of any SMO should be to ensure that IT Services are fit for purpose, fit for use, stable, secure, reliable, and fully support the business customer and IT objectives.

The SMO is chartered at the highest organizational level. When considering the establishment of an SMO, there is no single solution; rather, it must fit the needs of the organization in alignment with the organization's business objectives. This leads to a great point in using this Framework and implementing an ITSM Program.

> Success or failure depends on how well the Framework is adapted to your organization. Each organization, private or public sector, has a unique culture with unique business drivers and desired objectives. Adapt this Framework to your organization – not the other way around – for best results.

The purpose of the SMO is to coordinate and govern the development and execution of a customer-centric, enterprise-wide approach to ITSM - one that drives improved Service quality and performance to support the business customers' strategic goals and initiatives through standardized ITSM processes.

Typically, SMOs are charged with some or all the following responsibilities:

- Coordinating the development and maintenance of a business-aligned IT Service Catalog and associated Service Level targets
- Coordinating and supporting the execution of ITSM processes
- Providing shared Services, such as ITSM tools or Service reporting
- Establishing an ITSM process architecture with appropriate implementation guidance
- Prioritizing and aligning ITSM development and improvement activities
- Managing organizational change while implementing a Service Management culture
- Measuring and demonstrating ITSM value to business stakeholders
- Championing Continual Service Improvement (CSI) and coordinating improvements

> The SMO is charged with overseeing and aligning organizational ITSM resources, capabilities, and initiatives across the enterprise in support of organizational goals and objectives. The SMO must have focused goals.

Below are areas that should be addressed (the list is not exhaustive):

1. **Ensure alignment between IT Services and business objectives:** Realize direct relationships between the IT Services delivered and the requirements of end-users to ensure IT delivers defined and measurable value to the supported business

2. **Drive efficiency and effectiveness across the IT organization and support of the business customers' goals and objectives:** Advocate defined, measurable IT Services and standardized Service Management

3. **Coordinate the development and improvement of ITSM competencies:** Establish consistent visibility into the business perspective of IT performance, with the mechanisms in place to inform future IT investment decisions to improve overall

Implementing and Improving ITSM

The SMO executive leader is typically appointed from the sponsoring authority. The leader functions within the scope of delegated authority to ensure that resources are available for each SMO practice area. The number of resources will vary within each organization, and in some cases, individuals may perform roles in more than one practice area. However, it is ideal for each practice area to be led by one or more personnel who lead the Subject Matter Experts (SMEs) assigned to the practice area. A sample of practice areas and their nominal descriptions are listed below even though organizations differ greatly in size and structure.

Practice Area	Description
SMO Leadership	<ul><li>Serves as the facilitator for</li><li>Manages stakeholder communications (plans, strategies, and channels) to include governance bodies.</li><li>Develops training and awareness programs.</li><li>Facilitates mentoring and training of ITSM teams.</li><li>Directs and coordinates SMO practice area activities in support of stakeholder requirements.</li><li>Execute Organizational Change Management (OCM) activities and champion the ITSM value proposition across the enterprise.</li></ul>

Practice Area	Description
IT Process Architecture	- Develops, sustains, and improves enterprise ITSM process architecture. - Defines minimum standards - Conducts process architecture reviews with ITSM design and implementation teams. - Facilitates integration and prioritization of ITSM initiatives through an Implementation Roadmap, and supports integration of processes throughout the ITSM Service Lifecycle. - Supports the provisioning of ITSM tools (i.e., ITSM toolset, CMDB and monitoring capabilities) and a shared reporting service.
Continual Service Improvement	- Provides guidance related to quality and capability improvement methods, process capability assessments, and other disciplines related to Service Management in segmented environments.

> As with every other part of an ITSM Program, it is vital to develop clarity regarding roles within the SMO. Well-defined leadership structures and reporting relationships are critical in seamlessly integrating the SMO into organizational leadership of the enterprise. Primary, cross-functional, and supporting roles should be defined relative to the organizational structure.

The SMO functions under the authority contained in the original charter that is authorized by the Chief Information Officer (CIO). In so doing, the SMO acts on behalf of the CIO to execute all ITSM activities defined within the scope of the charter.

The scope of activities for the SMO usually includes developing and managing the integrated ITSM process reference architecture and establishing a system to measure and monitor Service performance and quality. The SMO champions the identification and designation of ITSM Process and Service Owners and provides guidance and oversight to ensure consistency and alignment with the policies, standards, and guidance established by the office.

Key functions and associated responsibilities include, but are not limited to:

ITSM Strategy
- Establish an organizational ITSM strategy and roadmap
- Facilitate the identification and assignment of roles and responsibilities in support of the strategy

Governance
- Establish, execute, and refine governance over ITSM resources and efforts
- Conduct regularly scheduled SMO meetings with assigned members to provide direction, facilitate decision-making, receive exceptions and proposals, and communicate status
- Identify, assign, and delegate authority to subordinate ITSM governance boards, working groups, and

Strategic Communications
- Manage Organizational Change Management through communications with members, stakeholders, and governance bodies
- ITSM advocacy support through the development of training and awareness program and facilitate mentoring and training of ITSM teams
- Champion the ITSM value proposition across the organization

Service Quality	Establish a Service Quality Management approach that defines the plan and methodology for achieving service quality for all Services and processes
	Develop and execute Service quality and performance standards to monitor and report the health of the IT Services
	Coordinate or conduct assessments of processes and Services
	Oversee and support execution of CSI efforts in compliance with process architecture quality, and performance management
Process Architecture	Develop enterprise ITSM process reference architecture, define minimum standards, and conventions for that process architecture
	Conduct process architecture reviews with ITSM design, implementation teams
	Facilitate integration and prioritization of ITSM initiatives via an Implementation Roadmap, and ensure interoperability
	Support the selection and maintenance of ITSM tools, such as ticket tracking systems, infrastructure and application management and monitoring tools
	Provide reporting on the health and performance of IT processes and services which may include issue resolution if the issue brought forth is a process issue
	Provide templates to promote standardization and consistency in support of ITSM efforts

Implementing and Improving ITSM

Core messages consist of general information that establishes a foundation of understanding to include ITSM objectives, and how they will impact stakeholders:

- ITSMO strategic vision and mission statement
- SMO approach to IT Service Management initiatives and training
- SMO Service Portfolio and Service Catalog

Key themes are critical advisory messages communicated to stakeholders, or targeted segments, to provide additional detail about ongoing initiatives and how they affect the community:

- ITSM strategy development
- SMO Governance of enterprise IT
- Training for Service Owners, Service
- Process Owners and Process Managers
- Strategic Communication of the ITSM value
- IT Service Quality capabilities
- ITSM process reference architecture establishment, updates, policies and standards

As with all other types of organizational change, the SMO should develop a comprehensive communications plan that addresses baseline messaging for internal and external stakeholders. All SMO communications should be predefined and crafted to ensure consistency and standardization of informational products. The messaging should be divided into two distinct groups.

The IT process architecture practice develops, sustains, and improves the organization's ITSM process reference architecture to enable and maintain the alignment of the IT processes and Services in support of the organization's business customers. The practice defines minimum standards and conventions for that process architecture and conducts process architecture reviews with ITSM design and implementation teams. It also facilitates integration and prioritization of ITSM initiatives through an Implementation Roadmap and supports integration of the processes throughout the ITSM Lifecycle. This architecture allows for the assessment and integration of new products and services required to improve the services offered to the business customer.

The Continual Service Improvement (CSI) practice area supports a Service Quality Management approach as well as IT Service Performance Management. This practice supports the organization in seeking to benchmark Services and implement Service improvement.

> The Service Management adage *"if you can't measure it, you can't manage it"* becomes apparent when delivering Services to the customer.

This practice area helps stakeholders define, gather, and analyze appropriate measures and metrics that enable continual process and Service improvement for their respective process or Service area. CSI is an ongoing effort as Service Quality improvements should never end. Service Quality is a measure of how well the Service delivered meets customer expectations. Service Quality and Performance Management foster an enterprise approach to govern Service quality measurement. This includes evaluating, directing, and monitoring Service quality measurement methods, approaches, techniques, and results along with recommending corrective actions. Evaluation activities ensure that there is an effective quality management approach and that measurements support customer and stakeholder requirements to include the identification of measurement gaps and a plan of action for closing those gaps to support decision-making.

Chapter 3 – Process Control

To enhance effectiveness and control of process policies, standards, process activities, performance measures, and overall process improvement, every process must include three common control activities. These common control activities provide a standard approach to process monitoring, reporting, and evaluating process performance and effectiveness and provide the process owner with built-in process governance and Continual Service Improvement for every process.

The establishment of a Process Framework consists of three steps, all common process control activities, shown below:

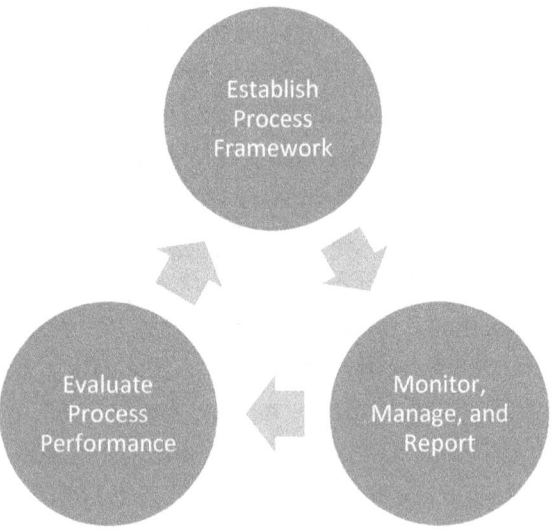

The Establish Process Framework activity is always the first activity. While more activities may be added, the Monitor, Manage, and Report activity is the next-to-last activity. The Evaluate Process Performance activity is the last activity. These control activities ensure a Continual Service Improvement loop in every process.

Establish Process Framework - This activity defines all direction, guidance, policies, and procedures for how to perform the process. This information is reviewed in the Evaluate Process Performance activity, which generates recommendations for changes and improvements to the process framework. The process framework is a collection of information, not necessarily a single document, which includes:

1. Process purpose, scope, goals, capabilities, and outcomes
2. Process policies, standards, and conceptual models
3. Process data requirements
4. Roles and responsibilities
5. Organizational responsibilities
6. Detailed procedures and best practices, including, but not limited to:
    a. Interfaces with other processes and programs
    b. Measurements and controls
    c. Tool requirements

7. The following tasks are performed in this activity:
    a. Review Process Evaluation Recommendations
    b. Specify Process Purpose, Scope, Goals, and Capabilities
    c. Define Process Policies, Standards, and Conceptual Models
    d. Determine Process Data Requirements
    e. Identify Process Roles and Responsibilities
    f. Assign Process Responsibilities to Organizations
    g. Determine Process Procedures
    h. Determine Process Relationships to Other Processes
    i. Define Measurements and Controls
    j. Determine Technology Needs
    k. Create Project Proposals
    l. Communicate and Deploy Framework

Monitor, Manage, and Support - Activities are monitored to determine whether suitable progress is being made. Results pointing to any need for leadership intervention are reported. The Process Manager is continually monitoring the normal work of the process.

The following tasks are performed in this activity:
1. Review progress of process maturity
2. Identify items of interest
3. Record and report findings
4. Perform/assign corrective action(s)

Evaluate Performance - This activity describes tasks required to assess how well the process is executed and recommends improvements to the process framework. It includes the capture of information on the relationship with other Services in the Service Lifecycle and/or process areas, and the suitability of procedures and training necessary to ensure continued success. This is an internally retained activity and provides for a continuous improvement loop ensuring that the process remains fit for purpose and identifies where changes to the process might be required. Evaluating process performance is a facet of Continual Service Improvement. The following tasks are performed in this activity:
1. Collect Feedback from Stakeholders
2. Produce Process Measurements
3. Research Trends and Best Practices
4. Collect Evaluation Results
5. Produce Gap Analysis
6. Recommend Initiatives
7. Complete Evaluation
8. Communicate to Stakeholders

Chapter 4 – Roles and Responsibilities

To best understand this key ITSM area, let's begin this chapter with definitions. Roles are a set of responsibilities, behaviors, activities, and authorities granted to a person or team. One person or team may have multiple roles. In this section, some potential key roles are defined. Below are the basic roles and a description of each:

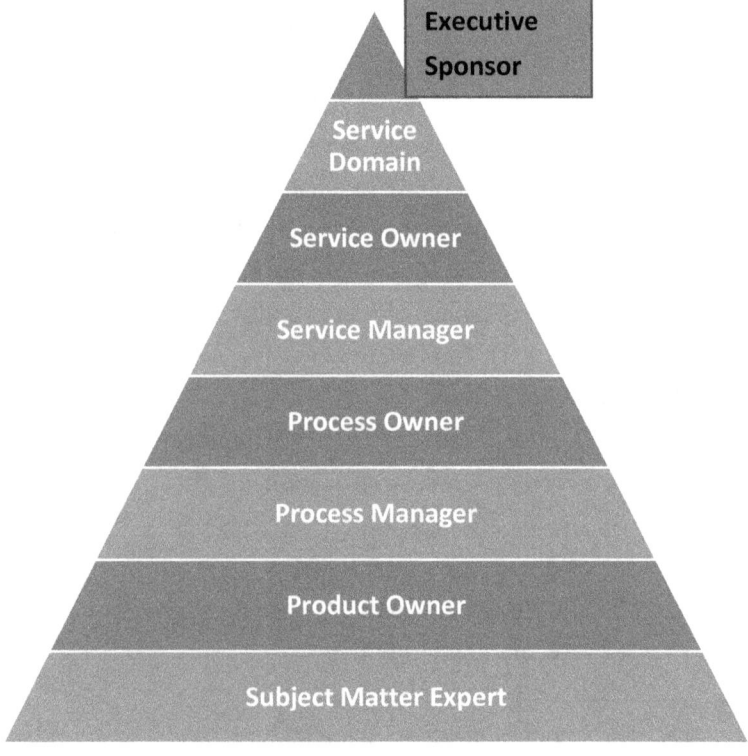

Most processes are made up of functions and roles. Functions are groups of people with specialized skills and responsibilities who are self-contained and develop a body of knowledge based on their expertise and experience. Roles are used to assign activities and responsibilities within a process. These are roles played within a process, not necessarily the job titles of the individuals. For example, a person may be an application analyst who plays the Problem Management Process Manager role.

> ITIL© utilizes the **RACI models to illustrate the roles and responsibilities within a Service or process.** Building an RACI model is easy and straightforward. First, identify and document the activities and processes. Then, identify and define the functional roles. Conduct a meeting to assign roles. Identify gaps and overlaps. Lastly, review and adjust regularly.

a). Executive Sponsor - The Executive Sponsor is accountable for the framework implementation and is responsible for securing spending authority and resources. The Executive Sponsor is a vocal and visible champion who legitimizes goals and an objective, keeps apprised of major activities, is the ultimate decision-maker, has final approval of all scope changes, and signs off on approvals to proceed to each succeeding phase.

b). Service Domain Owner - The primary responsibility of the Service Domain Owner is to ensure that the processes within the Service Domain provide support to the Service Owners, who have accountability for the Services that are provided. The Service Domain Owner is accountable for all the processes in the Service Domain, for the interfaces and process interdependencies, and for process maturity levels. The Service Domain Owner ensures proper resourcing, the appointment of Process Owners, and the strategy for each Service Domain. The Service Domain Owners work jointly to ensure proper handoffs between the Service Domains. The Service Domain Owners represent their Service Domain on the upper governance boards while establishing governance boards to handle Domain-specific matters related to policy standards.

c). Service Owner - The Service Owner is accountable for one or more Services throughout the entire Service Lifecycle, regardless of where the technology components, processes or professional capabilities reside. This includes the synchronization of resources that support the Service including resources that are located out of the Service Owner's organizational control. The Service Owner is responsible for Continual Service Improvement and the management of change affecting the Services under their care and is a primary stakeholder in all the underlying IT processes that enable or support the Service they represent.

Implementing and Improving ITSM

This role has the authority and responsibilities to ensure that activities are performed to identify, document, and fulfill Service requirements. The Service Owner is also responsible for ensuring that the following controls are built into the Service during Service Design:

- Business customer requirements
- Operational requirements related to an organization's Event Management, Business Continuity, Knowledge Management, and training
- Requirements for both normal operations and after an Incident
- Cyber-Security requirements for required stakeholders
- Auditing requirements, both financial and for Service Level Agreement (SLA) compliance
- Any required information-sharing interface points

d). Service Manager - This role is responsible for managing the end-to-end lifecycle of one or more IT Services. The Service Manager provides leadership on the development of the business case and process architecture as well as the Service deployment and lifecycle management schedules and performs Service cost management activities in close partnership with other organizations such as operations, engineering, and finance. The Service Manager is also responsible for the controls built into the Service.

e). Process Owner - The person fulfilling this role is accountable for ensuring that the process is fit for purpose, is being performed as agreed and documented, and is meeting the objectives of the process definition. The Process Owner's responsibilities include sponsorship, design, change management of the process documentation, and continual improvement of the process and its metrics. There is only one Process Owner per process. This role can be assigned to the same person who carries out the process manager role, but the two roles may be separate in larger organizations. The Process Owner has the following responsibilities:

- Accountable for the process design
- Establish a team to design and define the enterprise process
- Ensure that the process is "Fit for Purpose"
- Document and publicize the process
- Define appropriate standards to be employed throughout the process
- Review integration issues between the various processes
- Ensure appropriate resourcing to implement the process

- Ensure that all relevant staff have the required technical and business understanding, knowledge, and training in the process and understand their role in the process
- Define KPIs to evaluate the effectiveness and efficiency of the process and design reporting specifications
- Periodically audit the process to ensure compliance with policy and standards
- Review opportunities for process enhancements and for improving the efficiency and effectiveness of the process
- Attend top-level leadership meetings to assess and represent the process requirements and provide management information

f). Process Manager - In matters that pertain to the process, the Process Manager is answerable to the Process Owner and performs the day-to-day operational and managerial tasks demanded by the process activities. While there should be one Process Owner for each process, there may be multiple Process Managers for that same process. The Process Manager does not necessarily fall within the Process Owner's organizational chain of command. In addition, the Process Manager has the following responsibilities:

- Ensure that the process is carried out correctly
- Manage resources assigned to the process team
- Provide management and other processes with strategic decision-making information related to the process
- Monitor the process, using qualitative and quantitative Key Performance Indicators (KPIs), and make recommendations for improvement
- Play a key role in developing requirements for the process tools
- Function as an escalation point for questions related to the process
- Identify training requirements for support staff and ensure that proper training is provided to meet the requirements
- Provide metrics and reports to IT leadership and business customers in accordance with outlined procedures and agreements

g). Product Owner - This role is responsible for overseeing the end-to-end lifecycle of one or more IT products. The Product Owner ensures that the product(s) are fit for purpose and meet requirements of all associated services.

h). **Subject Matter Expert (SME)** - SMEs support various aspects of ITSM including the development, implementation, management, and improvement of ITSM processes to facilitate the delivery of IT Services. The SME further provides direction and support to integrate the process with other Service-supporting processes.

i). **Miscellaneous Roles** - An organization may create any role that is necessary to support its ITSM efforts. Other key participants/roles in the process implementation effort are:

1. **Senior Leaders/Directors** – Depending on how an organization is structured, there may be senior leaders who are not Service Domain Owners but have demonstrated an interest in the outcome of process implementation and are ultimately responsible for securing spending authority and/or providing resources.
2. **Process Design Team or Core Team** – Process definition and development will require a cohesive team of process experts and SMEs that come together until the process is implemented or objectives are met, and then the team is disbanded. This team may include roles such as Process Developer, Process Analyst, and Process Architect.

3. **Dedicated Members** – Points of contact from the various stakeholder organizations and field offices who participate in the implementation. They act as the contact point for the implementation team from the affected groups and provide status information to the management of their respective organizations.
4. **Stakeholders** – Individuals from the organization who have a stake in the process implementation and business partners whose support is needed during the implementation.
5. **Financial Support** – Someone who understands the funding required for resourcing the implementation project and understands Return on Investment (ROI) and Total Cost of Ownership (TCO) concepts, both in general and in context to the organization.
6. **Training Coordination** –As training programs are developed, it will be necessary to have a training expert to assist in ensuring that employees who take the training can be credited for their time and that leadership has a method for tracking the training attendance.
7. **Service Risk Management Owner** – This role is specifically recognized to identify and manage risk from a global or enterprise level. This role would be accountable for risk activities in all the Service Lifecycle domains and would work with each of the Service Domains to prepare and plan for risks.

8. **Service Strategy (or other Service Domain) Risk Management Analyst** – This role will support a single assigned Service Domain and be accountable to either the assigned Service Domain Owner or, if Service Risk Management has its own domain, to the Service Risk Management Domain Owner. This role looks at risks in every process in which he or she is assigned and assists with proactively planning and mitigating risks.

The RACI model described below is adaptable to every organization and establishes expectations for each role.

Roles

Responsible - Person or persons who executes the process and activities. Can have multiple "Rs"

Accountable - Accountable person who owns the quality and performance of the process. Person who has authority for decisions and owns the end-to-end quality and design. Can be only one

Consulted - This is someone we have a two-way conversation conveying input and knowledge. This role is usually a subject matter expert

Informed - These are people who receive information regarding the execution and quality of the process.

A few things to be mindful of in terms of the RACI model:
- As stated above, there can be only one person accountable and multiple people can play the other three roles.
- Responsibility and Accountability are difficult to delegate down in an organization.
- It is best not to match processes and activities with organizational departments. This maintains the integrity of the silos the processes are trying to cross.
- Combining responsibility for more than one process can create conflict. As we mentioned above, the outputs of one process are the inputs to another process.

Below is the RACI for a deli:

Activity| Role	ITSM Deli Owner	Chef	Sous Chef	Waiter	Accountant
Menu Creation	A	R	I	I	
Take Order	A	I	I	R	
Prepare Order	A	R	R	C/I	
Serve Food	A	C	C	R	
Check Out	A			R	
Financial Analysis	R				A

Implementing and Improving ITSM

The above RACI is very simplistic but gives a basic understanding of how an RACI may be used to help everyone understand their roles and what is expected in those roles. When we apply the RACI to an IT Service Provider organization, it may look like the below:

Role	Notes
Process Owner	The strategic role accountable for the Process. This role: • Is accountable to senior leadership for the proper design, execution, and improvement of the process • Ensures that the process is being carried out, but does not run the day-to-day operation of the process • Receives regular updates concerning the performance of the process and represents this process concerning all decisions being made by senior management

Implementing and Improving ITSM

Process Manager	This is a tactical role and: • Runs the day-to-day operation of the process • Takes overall direction from the Process Owner • Oversees the direction and operation of the process • Provides appropriate reporting to interested parties
Process Developer	This role is responsible for coordinated process development: • Create Charter • Form Process Development Team • Develop Communication Plan • Develop High-level Design • Develop Detailed Design • Develop Transition Plan for Implementation
Process Analyst	This role is responsible for: • Process Execution • Completing required training • Supporting Process development, design, and implementation • Implementing process • Providing continuous process improvement support

Chapter 5 – Process Design Steps

The following steps apply to all processes:

1. Define Scope and Activities
    - Determine and document business objectives, problem(s) to solve, and scope of project
    - Gain necessary agreement or authority to proceed
    - Determine relevance, impact, and priority of IT strategic plans and related policies
    - Create Implementation Road Map
2. Validate the Current Environment
    - Collect existing process documentation
    - Identify existing roles and responsibilities
    - Document current, "As-Is" process environment
    - Identify issues for "Quick Wins"
    - Identify supporting tools
    - Document tool and process gaps

3. Develop High-Level Process Definition
   - Identify CSFs and high-level KPIs
   - Document high-level process definitions
   - Define high-level process inputs and outputs
4. Define Roles and Responsibilities
   - Identify skills and knowledge level
   - Create RACI matrix mapping activities to process roles
   - Develop cross-functional relationships
5. Document Detailed Work Flow for each High-Level Activity
   - Document detailed procedures for each high-level activity
   - Create communications plan
6. Build the Process
   - Document "To-Be" process environment
   - Create workflow
   - Incorporate common Process Control activities
   - Define inputs and outputs at a detailed level
   - Incorporate control information
   - Determine appropriate metrics
   - Document tool requirements, including interfaces to other processes

7. Develop Metrics and Supporting Metrics
    - Review which metrics to collect on a regular schedule. There needs to be a reason and a decision for each metric collected. If there are no longer decisions being made on metrics, do not use resources in collecting and analyzing the metric
    - Metrics should show a change in percentages, not simply a change in number -- i.e., percentage of incidents assigned incorrectly is more effective than number of incidents assigned incorrectly
8. Define and Document Communications Plan, Knowledge Transfer, and Training Requirements
    - Develop Communications Plan
    - Develop Training Program
    - Deliver Process and Tool Training
9. Identify and Implement Quick Wins

During a process-improvement initiative, one must balance the need for a stronger process foundation with the need to demonstrate more immediate value from the process. The result of this analysis is known as Quick Wins. An immediate focus on Quick Wins helps engage key members of the organization to improve the process. Their involvement is recognized by others and establishes momentum for additional improvements that may require more time and commitment.

The key is to begin building sponsorship at all levels of the organization by demonstrating real benefits that add real value.

Quick Wins have common characteristics:
- Lower level of effort than other initiatives while still adding value in a relatively short period of time
- Important to the overall process-improvement effort
- Eases organizational issues

10. Finalize Process Guide
- Use the appropriate Process Guide template to produce and publish guide
- Content will have been created throughout the process design
- Post the Process Guide and communicate its existence and location
- Incorporate version control per organization rules

Section Two

Chapter 6 – Service Strategy

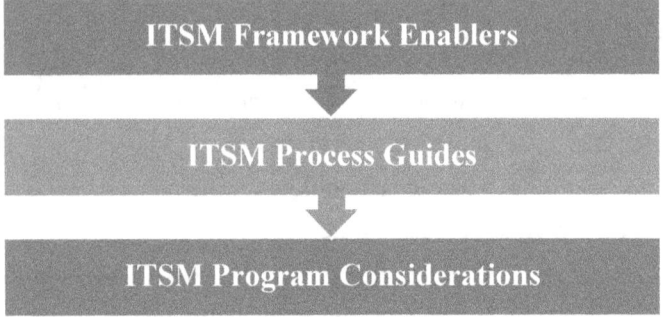

At the center of the Service Lifecycle is Service Strategy. This is where organizational objectives and business customer needs are aligned. Service Strategy ensures that the organization can identify, understand, and control the costs and risks associated with the Service Portfolio, and has the foundation for operational effectiveness and quality performance.

The processes in Service Strategy provide guidance and direction to support the business customers and identify, select, and prioritize Service opportunities. A prime goal of Service Strategy is to understand why a Service is provided *before* deciding how to provide the Service.

Below is a deep dive into the key processes of Service Strategy, and the other Service Lifecycle stages subsequently.

1). **Business Relationship Management (BRM)**

Purpose - The purpose of Business Relationship Management (BRM) is to identify, monitor, and manage customer and stakeholder needs and expectations. Corrective actions are taken and implemented based on the Business Relationship data collected to meet and increase customer satisfaction.

Scope - BRM involves all business outcomes related to business customer engagements. This relationship covers the entire lifecycle of the Services offered, from the agreement to create a Service to the retirement or decommissioning of a Service. BRM and Service Level Management (SLM) are similar in that each has a high degree of business customer interaction. The specific difference is that BRM builds the relationships with business customers and SLM defines business customer requirements and negotiated Service Levels.

BRM understands the business objectives, as well as the environment in which the Services operate. This enables the Service Provider to identify and respond to the needs of the customer and manage business and stakeholder expectations of the Service Provider. Customer relationships, as a subset of the business relationship, are fostered and aligned to maximize customer satisfaction, value perception, and retention.

Process Benefits
- Facilitates ongoing communication with business customers
- Focus on the cultural aspects of business satisfaction
- Reduced breaches of Service Level Agreements
- Ability to anticipate business customer needs through the greater understanding of their goals and use of provided capabilities
- Increased trust because of the partnership established between the Service Provider and business customer

Implementing and Improving ITSM

Expected Outcomes
- A strategy for defining and maintaining business relations between the business customer and the provider is defined and implemented
- Awareness of business customers, their needs, and major changes are maintained
- Service performance status and reports are monitored for potential business relationship or customer satisfaction impacts and improvements
- Business partner satisfaction is monitored, measured, and reported
- Approved actions to maintain or improve business satisfaction are implemented
- Contractual disputes are resolved
- Service complaints (and compliments) are recorded, investigated, acted upon, reported, formally closed, and, when necessary, escalated

Process Activities

[BRM1] Establish Business Relationship Management Framework

This activity defines all direction, guidance, policies, and procedures for how the process will be performed. All of this is collectively referred to as the "BRM process framework" and is used as reference information for all other activities. This information is reviewed in the Evaluate Process Performance activity, which generates recommendations for changes and improvements to the BRM process framework.

[BRM2] Capture Business Relationship Data

This activity involves gathering business relationship data, including the identification of business customers and establishing points of contact. Data is also gathered on data points identified such as technology metrics (utilization, performance, and availability), process metrics (Service Level Agreements (SLA), Key Performance Indicators (KPIs), and activity metrics for the Service Level Management process), customer satisfaction metrics and Service execution metrics. This activity gathers only information needed for analysis. As the organization is collecting the data, both reactive and proactive methods are employed.

[BRM3] Analyze Business Relationship Data

Analysis of the data collected identifies:
- Results for the current reporting period regarding Business Relationships
- Business Relationship trending
- Root causes for underlying customer satisfaction issues

[BRM4] Manage Business Relationship Issue Resolution

Analysis results are used to create action plans that address issues and provide the status of issue resolutions to stakeholders. The notification and communication plan are incorporated into this activity to apprise senior leadership.

[BRM5] Assess Business Relationship Patterns

This activity performs trending analysis of satisfaction data. Its purpose is to identify the underlying cause of trends; negative and positive. Once identified, the issue is communicated and assigned to appropriate resolution plan.

[BRM6] Monitor, Manage and Report Business Relationship Management

This activity supports continuous monitoring and analysis of operational results data and comparison with Service achievement reporting to identify Business Relationship Management trends and issues. Business Relationship Management information is used to generate detailed Service component reporting as well as provide perspective on overall Service availability. All Business Relationship Management activity is monitored to determine whether suitable progress is being made. Unsatisfactory results are reported and may result in actions taken to address any issues.

[BRM7] Evaluate Business Relationship Management Performance

This activity describes the tasks required to assess the efficiency and effectiveness of the Business Relationship Management process. It includes the capture of information on records, the relationship with other process areas, and the suitability of procedures and training. It is used as a basis to ensure that the Business Relationship Management process remains fit for purpose and identifies where changes to the process might be required.

2). **Demand Management (DM)**

Purpose - The purpose of Demand Management is to provide an understanding of the business customer demand for a specific Service and to plan early for provisioning of capacity and other aspects of support for the Service. This process may influence business customer demand for Services and seeks to proactively understand the business customer workload (demand) with the available resources (supply) through analysis, trending, and forecasting.

Scope - The Demand Management process seeks to understand the expected business behavior of all demand resources across the enterprise at the individual user level and aggregated level, and to represent the overall impact on IT. The Demand Management process translates demand from business requirements in IT Service terms (i.e. User Profiles). It identifies gaps and misalignment between demand and supply, and proposes policies and incentives designed to minimize or close gaps. This is beneficial to planning IT capacity and other resources as required.

Process Benefits

- Quicker reaction to changing needs
- Leads to more accurate cost information to support IT investment decisions and to determine cost of ownership for ongoing Services
- Enables planning a more efficient use of IT resources
- Service demand is a key factor in prioritization of resources
- Proactive contingency plans are in place for demand variances

Expected Outcomes

- Improved IT flexibility in response to dynamic business (business customer, supplier, environmental, etc.) changes through a structured approach to evaluating strategic and financial impact of Service demand
- Improved ability to support SLAs and OLAs because demand, and its impact, is better understood
- Improved decision-making through a constantly evolving knowledge base, designed to consistently re-evaluate proposed initiatives
- Improved Capacity Planning efforts through improved forecasting of demand
- Optimized IT resource utilization
- Improved visibility of ITSM operations

Process Activities

[DM1] Establish Demand Management Framework

This activity defines all direction, guidance, policies, and procedures for how the process will be performed. All of this is collectively referred to as the "Demand Management process framework" and is used as reference information for all other activities. This information is reviewed in the Evaluate Process Performance activity, which generates recommendations for changes and improvements to the Demand Management process framework.

[DM2] Evaluate Demand Requirements

This activity determines the means for analyzing business demand. This is important because it establishes the data collection requirements from processes that provide raw data via Knowledge Management (e.g. Request Fulfillment and Capacity Management). The execution of this activity is advised to properly establish strategy prior to the collection, analysis, and subsequent decisions that occur in Demand Management.

[DM3] Gather Demand and Usage Data

This activity collects and consolidates demand data from multiple sources for further analysis. A comprehensive analysis of demand is used for demand forecasting and initiative evaluation.

[DM4] Identify Patterns of Business Activity and User Profiles

In this activity, patterns of business behaviors are evaluated and used to synchronize consumption (demand) with capacity (supply) of IT resources. Incoming data and known upcoming initiatives from Service Portfolio Management are helpful to determine requirements for the Demand Management process.

[DM5] Develop Demand Forecast

This activity uses the Service demand baseline and collected data along with aggregated historical data to generate a demand forecast. This forecast will provide insight into upcoming demand requirements, including expected high/low demand periods.

[DM6] Plan and Implement Demand Management Initiatives

This activity uses Demand Forecast information to predict misalignment between demand and supply of IT resources and Services. It creates a strategy to realign resources and Services through policy, incentives, and/or IT resource investment. When a decision to shape demand through incentives is made, analysis is performed to shape demand through methods such as Differential Pricing, Off-Peak Pricing, Volume Discounts, Tiered Service Levels, etc. This activity concludes with the formulation and communication of a prioritized set of recommendations (e.g. plans of action, investment recommendations, etc.)

[DM7] Monitor, Manage and Report Demand Management

In this activity, all Demand Management activity is monitored to determine whether suitable progress is made. Unsatisfactory results are reported and may result in actions taken to address any issues.

[DM8] Evaluate Demand Management Performance

This activity describes the tasks required to assess the efficiency and effectiveness of the Demand Management process. It includes the capture of information, relationship with other process areas, and suitability of procedures and training. It is used as a basis to ensure that the Demand Management process remains fit for purpose and identifies where changes to the process might be required.

3). **Financial Management (FM)**

Purpose - The purpose of the Financial Management process is to control budgeting, accounting, and chargeback for Service provision. From the IT standpoint, Financial Management secures funding to acquire and maintain the enterprise architecture. Finally, Financial Management should provide transparency into the spending and cost recovery for Services provided in the IT environment.

Scope - The scope of FM for IT Services covers aspects of three sub-processes associated with overall process budgeting, accounting, and charging. Budgeting tasks include predicting, controlling expenditures and monitoring budgetary adjustments. Accounting identifies the costs of delivering IT Services, compares those costs with budgeted costs, and manages variance from the budget.

All accounting practices must be aligned to the wider accountancy practices of the whole of the service provider's organization. If applicable, a charging system is developed to recover the cost of IT provision.

Process Benefits

- Increased confidence in setting and managing budgets
- Accurate cost information to support IT investment decisions and determining cost of ownership for ongoing Services
- IT is understood in concepts of "Return on Investment" (ROI) and "Total Cost of Ownership" (TCO) as related to Services provided
- Cost and expenditures are better understood by the IT staff
- Controls demonstrating compliance with congressional mandates are recognized and built in during strategy
- Increased focus on current and future IT Service areas

Expected Outcomes

- Cost estimates are developed
- Results from cost estimates are used to produce budgets
- Deviations from the budget and costs are controlled
- Corrective actions are taken to resolve deviations from the budget
- Charging is implemented to recover the cost of IT provision, if applicable
- Deviations from the budget and costs are communicated to affected parties

Process Activities

[FM1] Establish Financial Management for IT Services Framework

This activity defines all direction, guidance, policies, and procedures for how the process will be performed. All of this is collectively referred to as the "Financial Management process framework" and is used as reference information for all other activities. This information is reviewed in the Evaluate Process Performance activity, which generates recommendations for changes and improvements to the Financial Management process framework.

[FM2] Perform Financial Modeling

Financial modeling determines likely financial outcomes for a wide range of propositions, whether limited to management of IT finances or to proposals relating to the business, infrastructure, Service variations or any other consideration requiring cost-benefit analysis. Requests will differ in some ways and require innovative modeling approaches. For example: Service valuation, Demand modeling, Service investment analysis and Variable cost dynamics.

[FM3] Plan and Control Budgets

Planning and controlling IT Service budgets provides for better cost accountability and more accurate forecasting of future budget requirements. If budget is exceeded early warnings are given.

[FM4] Perform Financial Accounting

Financial Accounting determines costs incurred to provide IT Services and provides high-level analysis of those costs and the value provided by the expenditure. The goal of financial accounting is to understand what drives IT costs and whether IT delivers good value for the money invested. As a result, Financial Accounting aids investment and renewal decisions, identifies poor value for money and costs of changes, and performs Return on Investment (ROI) and cost-benefit analysis.

[FM5] Audit Financials

The purpose of the Audit Financials activity is to confirm conformance to financial standards and best practices. Financial data is examined using defined criteria and guidelines.

[FM6] Monitor, Manage and Report Financial Management for IT Services

In this activity, all process activities are monitored on an ongoing basis to ensure that suitable progress is made. Gaps are reported and may result in corrective modifications to the processes. This process also manages requests for information and status.

[FM7] Evaluate Financial Management for IT Services Performance

The purpose of this activity is to evaluate the performance of the Financial Management for IT Services process and identify improvement areas to the overall process. Continuous Service Improvement considerations include reviews of foundations and interfaces, all activities within the process, and adaptability of the process and the roles and responsibilities assigned. FM is also evaluated against goals and measures to quantify its influence on overall IT improvements. Improvements include insights and lessons learned from observation and analysis of activity accomplishments and results.

4). Service Portfolio Management (SPM)

Purpose - The purpose of the Service Portfolio Management Process is to evaluate and prioritize Service investments proposals to ensure value to the business. It is involved in the entire lifecycle of the Service, from the time Service is requested, until it is decided that the Service will be discontinued and decommissioned. Service Portfolio Management ensures that the right set of Services is offered to meet the business at the appropriate cost level. It is the decision framework that facilitates the decision-making process regarding what Services are offered to meet business customer needs. A Service Portfolio is different than an IT Portfolio. It may be implemented as a part of an IT Portfolio, Project Portfolio or Enterprise Portfolio. The focus of the Service Portfolio Management process is on Service offerings to the business customer.

Scope - The scope of Service Portfolio Management encompasses all IT-related Services offered and may reside in one of three Service Lifecycle phases:

- **Service Pipeline** – Services under consideration for investment or residing in Service Design or Service Transition
- **Service Catalog** – Services that are currently available and can be browsed by the business partner base, normally considered operational and candidates for Continual Service Improvement
- **Retired** – Services that are no longer deployed and are unavailable to the business customer

Process Benefits

- Aligned and prioritized Services
- Focus on managed Services increases efficiency in bringing new Services to realization
- Allocation of resources for changes and additions to the Service Portfolio are business based
- Risk Assessment for creating Services are handled in a consistent, measured process
- Services are continually and consistently evaluated for their value to the business

Expected Outcomes

- Aligned Service investment decisions with business and business customer needs
- Defined inventory of Services
- Validated portfolio data
- Maximized portfolio value
- Created/analyzed business cases
- Aligned and prioritized Services
- Balanced supply and demand
- Decisions are communicated, and resources are properly allocated
- Investments are prioritized and selected based on stakeholder goals and return on investment

Process Activities

[SPM1] Establish Service Portfolio Management Framework

This activity defines all direction, guidance, policies, and procedures for how the process will be performed. All of this is collectively referred to as the "Service Portfolio Management process framework" and is used as reference information for all other activities. This information is reviewed in the Evaluate Process Performance activity, which generates recommendations for changes and improvements to the SPM process framework.

[SPM2] Create Initial Services Inventory

Collect data about all Services. This includes Services that are operational as well as Services currently in development. This activity should be done only once to create the Service Portfolio and the parameters around how Services are made visible to the customers or retired as no longer offered or supported. If the Service is currently offered, it belongs in the Service Catalog (a subset of the Service Portfolio) where it is visible to customers and under control of the Service Catalog Management Process.

If it is a Service that is currently in the design phase, it is in the Service Pipeline and under control of the SPM process to determine when the Service can be visible to customers via the Service Catalog. If the Service is retiring, SPM ensures appropriate activities and that customer visibility to the Service is removed from Service Catalog.

[SPM3] Define Service Analysis Objectives and Thresholds

Define Service analysis objectives and thresholds to develop a roadmap to identify how Services are assessed and moved through different stages of the Service Lifecycle. These objectives are used to assess candidate Services and include references to availability and capacity plans, financial constraints, customer satisfaction objectives, and other artifacts. Thresholds for moving a Service from the Service Pipeline to the Service Catalog are also determined.

[SPM4] Assess and Prioritize Service Proposals

Review Service proposals and determine which should be accepted for consideration. Develop or update the business case for each Service proposed, which may include the identification of the following:

- Key Performance Indicators (KPIs)
- anticipated user base and frequency
- alternative solutions
- technical scope
- financial metrics (ROI, TCO)
- benefit cost analysis
- intangibles
- major assumptions and constraints
- opportunity cost analysis
- gap analysis
- analysis of alternatives
- sensitivity analysis
- risk assessment
- impact analysis
- contingencies

Categorize and prioritize each Service or change to Service proposed.

[SPM5] Determine Service Approval

Review Service proposals and decide on approval to proceed. Request additional information for clarification as needed. Update the Service Portfolio accordingly.

[SPM6] Conduct Service Portfolio Review

Perform a comprehensive review of the Service Portfolio and evaluate Service balance and alignment. The review determines corrections to the mix of Services to better maximize Services offered. Monitor and evaluate to ensure that Service is within agreed cost, schedule, and scope constraints. Evaluate actual results against planned results.

[SPM7] Monitor, Manage and Report Service Portfolio Management

In this activity, all Service Portfolio Management activity is monitored to determine progress. Unsatisfactory results are reported and may result in actions taken to address any issues.

[SPM8] Evaluate Service Portfolio Management Performance

This activity describes tasks required to assess the efficiency and effectiveness of the Service Portfolio Management process. It includes the capture of information on records, the relationship with other process areas, and the suitability of procedures and training. It is used as a basis to ensure that the Service Portfolio Management process remains fit for purpose and identifies where changes to the process might be required.

Chapter 7 – Service Design

Service Design ensures that Services are designed to align and match current and future requirements. As a Service Lifecycle phase (and sometimes called a "Domain"), it controls planning and organizing resources, infrastructure, communications, and physical and logical components of Services to improve Service quality and the interaction and understanding between the Service Provider and its business partners. This culminates in a comprehensive Service Design Package (SDP).

The Service Lifecycle phase ensures that goals and objectives of Service Strategy are built and managed in line with the vision and business. Service Design relies heavily on Service Owners to understand requirements, needs, and Service behavior of business partners. It is accountable for changes to existing Services, creation of new Services, and management of the removal of existing Services. Service Design coordinates with Service Operations to ensure that the data necessary for monitoring and responding to service variances is built into every Service.

1). **Service Catalog Management (SCM)**

Purpose - The purpose of the Service Catalog Management process is to provide an authoritative source of consistent information on all available Services and to ensure that the information is accessible to those who are authorized to view it.

Service Catalog Management defines, collates and publishes approved descriptions, under change control, of all Services using terms aligned to the customer's view of Services and understandable by those without a detailed technical understanding.

Scope - The scope of Service Catalog Management is to provide and maintain accurate information on all active Services and all Services in transition to production. These Services may be represented individually, or as packages. Information about the Services includes Service definition, Service Levels, points of contact, ordering and Service Request information. SCM correlates closely with Service Portfolio Management regarding Service offering timelines, Service interfaces, and dependencies.

Process Benefits
- Provides business partners an automated interface to the "menu" of Services
- A process for maintaining the information for the Services provided in a controlled fashion
- Visibility of Services to assist in decisions

Expected Outcomes

- A single authoritative source of information on Services offered
- Accurate information on all operational Services and those about to be offered (details, status, interfaces, and dependencies) is maintained in the Service Catalog
- Views of the Service Catalog provide an understanding of Service definitions and use

Process Activities

[Step 1] Establish Service Catalog Management Framework

This activity defines all direction, guidance, policies, and procedures for how the process will be performed. All of this is collectively referred to as the "SCM process framework" and is used as reference information for all other activities. This information is reviewed in the Evaluate Process Performance activity, which generates recommendations for changes and improvements to the SCM process framework.

[Step 2] Define Service Catalog Requirements

This activity identifies all the requirements for a Service Catalog, including overall structure, content requirements, navigation, views for different user groups, etc. Requirements come from a variety of sources, including Service Portfolio Management, Service Level Management, user representatives and stakeholders.

[Step 3] Plan Service Catalog

After requirements are defined for the Service Catalog, this activity plans and designs the Service Catalog. This involves designing catalog appearance, structure, navigation, relationships and ensuring that the catalog is actionable.

[Step 4] Implement and Modify Service Catalog

The implementation and modification of the Service Catalog are carried out by this activity. This activity executes all tasks associated with catalog structure, appearance, navigation, and content. All modifications are approved before the catalog is published.

[Step 5] Publish Service Catalog

In this activity, a newly implemented or updated Service Catalog is published to authorized user groups.

[Step 6] Monitor, Manage and Report Service Catalog Management

This activity supports continuous monitoring and analysis of operational results data and comparison with Service achievement reporting to identify Service Catalog Management trends and issues. Service Catalog Management information is used to generate detailed Service component reporting as well as a perspective on overall Service availability.

[Step 7] Evaluate Service Catalog Management Performance

This activity describes the tasks required to assess the efficiency and effectiveness of the Service Catalog Management process. It includes the capture of information, relationship with other process areas, and suitability of procedures and training. It is used as a basis to ensure that the Service Catalog Management process remains fit for purpose and identifies where changes to the process might be required.

2). **Design Coordination (DC)**

Purpose - The purpose of Design Coordination is to ensure the consistent and effective design of new or changed IT Services as well as the retirement of IT Services. Design Coordination facilitates all Service Design objectives are met by providing a single-point-of-contact for the efforts in this lifecycle stage.

Scope - Design Coordination includes all new or changed Services that enter the design phase. This is primarily as part of a project and requires coordination with the Transition Planning and Support counterpart. Design Coordination will work with the Service Owner to ensure that all requirements are integrated into the new or updated Service Design Package (SDP) for each IT Service. Additionally, Design Coordination will work with the Service Owner to ensure that all security requirements are integrated into the new or updated SDP for each IT Service.

Process Benefits
- Reduced costs associated with reworking design
- Accountability for the Service Design Package (SDP)
- Ensured architectural consistency
- Consistent design approach and coordination of all design activities
- Ensures that all Service Models and Service solution designs conform to security policies

- Reusable design practice

Expected Outcomes

- Consistent approach to design of Services
- Delivered Services that are effective and efficient through coordination of all design activities
- Designed Services that can be easily and efficiently developed
- Overall improvement in the quality of IT Service within the imposed design constraints by reduction in rework once they have been transitioned into the live production environment
- Service Models and Service solution designs adhere to strategic, architectural, governance, and organizational requirements

Process Activities

[DC1] Establish Design Coordination Framework

This activity defines all direction, guidance, policies, and procedures for how the process will be performed. All of this is collectively referred to as the "Design Coordination process framework" and is used as reference information for all other activities. This information is reviewed in the Evaluate Process Performance activity, which generates recommendations for changes and improvements to the DC process framework.

[DC2] Plan Design Resources and Capabilities

The purpose of this activity is to coordinate and plan the resources, capabilities, standards, methods, techniques, technologies, and environments related to a specific SDP.

[DC3] Coordinate Design Activities

In this activity, the focus is on the coordination of all design activities across projects/changes and the management of schedules, resources, conflicts, suppliers and support teams as required.

[DC4] Manage Design Risks and Issues

In this activity, formal risk assessment and management techniques are used to manage risks associated with design activities and reduce the number of issues that can be attributed to poor design.

[DC5] Monitor, Manage and Report Design Coordination

This activity supports continuous monitoring and analysis of operational results data and comparison with Service achievement reporting to identify Design Coordination trends and issues. Design Coordination information is used to generate detailed Service component reporting as well as perspective on overall Service availability.

[DC6] Evaluate Design Coordination Performance

This activity describes tasks required to assess the efficiency and effectiveness of Design Coordination. It includes the capture of information on records, relationships with other process areas, and the suitability of procedures and training. It is used as a basis to ensure that the Design Coordination process remains fit for purpose and identifies where changes to the process might be required.

3). **Availability Management (AVM)**

Purpose - The purpose of the Availability Management process is to ensure that availability of approved IT resources to meet business requirements are consistently met or exceeded. Availability Management is concerned with meeting future availability needs of a new or expanding Service base and ensures that Services remain cost-effective.

Scope - Availability Management is responsible for safeguarding the interests of the stakeholders and interested parties by ensuring that approved Service Levels are met as defined in Service Level Agreements (SLAs). It includes defining, analyzing, planning, measuring and continually improving all aspects of IT resource availability. This process produces and maintains an up-to-date Availability Plan that reflects current and future needs.

Process Benefits

- Ensures that an Availability Plan is developed and is in alignment with business goals and agreements
- Resources are better utilized as Services are placed on infrastructure that is based on availability requirements
- The Availability Plan helps identify Service Availability issues prior to outages
- Business partner satisfaction rises as Availability increases and Incidents decrease
- SLAs are met regarding uptime and availability
- There is a quantitative approach and plan to address availability
- Services are designed and engineered to meet availability requirements
- Issues with availability are viewed holistically, not Service by Service
- An up-to-date Availability Plan mitigates risk when new Services are under consideration or a Service is deploying to an expanded user base
- AVM works in conjunction with Capacity Management (CapM), IT Service Continuity Management (ITSCM) and Information Security Management (ISM) – the "Warranty" processes for Service Warranty related to SLAs of a Service

Expected Outcomes

- Service Availability requirements are identified
- A Service Availability Plan is developed using Service Availability requirements
- Service Availability is tested against the Service Availability requirements to validate the plan
- Service Availability is monitored
- Underlying causes of unanticipated Service non-availability are identified and analyzed
- Corrective actions are taken to address identified underlying causes of non-availability
- Changes to Service Availability requirements are reflected in the Service Availability Plan

Process Activities

[AvM1] Establish Availability Management Framework

This activity defines all direction, guidance, policies, and procedures for how this process will be performed. All of this is collectively referred to as the "Availability Management process framework" and is used as reference information for all other activities. This information is reviewed in the Evaluate Process Performance activity, which generates recommendations for changes and improvements to the Availability Management process framework.

[AvM2] Monitor and Report Service Availability

As appropriate for the infrastructure environment, this activity monitors and reports on Service and network availability using defined toolsets. Service Availability measures the end-to-end availability of critical and noncritical Services provided. Network availability is defined as the percentage of time the network is capable of transmitting data as designed among users and to and from gateways to external networks and sites.

[AvM3] Collect and Analyze Monitoring Data

In this activity, Service Availability monitoring data is obtained and analyzed. The data comes from a variety of sources, including: Service Level monitoring data, Incident information and trends, Problems and Known Errors and Service testing data.

[AvM4] Assess Availability Risks

SLAs, OLAs, and UCs are reviewed for availability terms, conditions and targets, and availability-specific requirements. Availability requirements contribute key data to the Availability Plan. This activity assesses the impact of changes to Services, versus monitoring to mitigate risks of non-compliance to or negatively impacting SLAs.

[AvM5] Plan Availability for New and Changed Services

The Availability Manager receives approved (Request for Change) RFCs, workarounds, and fixes for availability incidents and problems. When new and changed Services are proposed, the Availability Management process will proactively adjust the Service Availability Plan, to allow for new SLAs and monitoring through the SLM process.

[AvM6] Create and Manage Availability Plan

This activity generates the Availability Plan that summarizes resource availability optimization decisions and commitments for the planning period. It includes availability profiles, targets, issues descriptions, and historical analyses of achievements regarding target summaries, and documents lessons learned. The Availability Plan is a comprehensive record of the approach and success in meeting the organization's expectations for IT resource availability.

[AvM7] Monitor, Manage and Report Availability Management

In this activity, Availability Management activities are monitored to determine whether suitable progress is being made. Results are reported, and unsatisfactory results may lead to a review of Availability Management actions. In addition, responses are provided to requests for information and status of the Availability Management process.

[AvM8] Evaluate Availability Management Performance

This activity describes the tasks required to assess the efficiency and effectiveness of the Availability Management process. It includes the capture of information on records, the relationship with other process areas, and the suitability of procedures and training. It is used as a basis to ensure that the Availability Management process remains fit for purpose and identifies where changes to the process might be required.

4). **Capacity Management (CapM)**

Purpose - The purpose of the Capacity Management process is to ensure that Service and component capacity meets current and future agreed requirements and performance levels. This information is maintained and updated in a Capacity Plan.

Scope - This process ensures that there are sufficient resources and capacity to meet current and future negotiated and approved requirements in a cost-effective and timely manner. Capacity Management ensures that proactive measures to improve Service performance are implemented wherever it is cost-justified. It maintains a balance between costs and capacity, supply and demand, and ensures that agreed performance levels are met. The scope includes almost all configuration items (CIs) and the following resources are taken into consideration:

- Computer hardware, network resources
- Software applications
- People
- Other environment resources like warming/cooling equipment, furniture for staff
  (Basically, anything that is a resource to the performance of the Services)

Process Benefits

- A Capacity Plan is developed
- Proactive management of capacity reduces performance and capacity related Incidents
- Uninterrupted availability and performance levels during peak periods
- Unnecessary expenses caused by "last minute" purchases or re-allocation of resources are avoided
- Mature Capacity Management is essential to cloud computing
- Infrastructure growth is planned to meet organization needs and demand
- Risk reduction in running the production environment
- Efficient infrastructure budget

Expected Outcomes

- Current and future capacity and performance requirements are identified and agreed
- A Capacity Plan is developed based on capacity and performance requirements
- Capacity is provided to meet current capacity and performance requirements

- Capacity usage is monitored, analyzed and performance is tuned

- Capacity is prepared to meet future capacity and performance needs
- Changes to capacity and performance are reflected in the Capacity Plan

Process Activities

[CapM1] Establish Capacity Management Framework

This activity defines all direction, guidance, policies, and procedures for how the process will be performed. All of this is collectively referred to as the "Capacity Management process framework" and is used as reference information for all other activities. This information is reviewed in the Evaluate Process Performance activity, which generates recommendations for making changes and improvements to the Capacity Management process framework.

[CapM2] Review Current Capacity and Performance

This activity invokes the monitoring of and a regular generation of reports on Service and component capacity and performance to ensure that Service performance meets or exceeds all performance targets. Reports from monitoring are - at a minimum - generated on a periodic basis or may be generated at customer or management request. The types of information monitored may be altered over time as determined by the Capacity Manager based on history, incidents, problems, services or management need. This information would generally be outlined in the Capacity Plan for the Service or the CapM Framework.

[CapM3] Assess, Agree, and Document New Requirements and Capacity

This activity monitors patterns of business and Service activity and Service Level plans through performance, utilization and throughput of IT Services and the supporting infrastructure, environmental, data, and application components. This activity involves the use of trending, forecasting, modeling techniques, and thresholds to plan upgrades, enhancements, and estimated future requirements. The current Capacity Plan is also considered before plans are solidified to procure additional capacity.

[CapM4] Improve Current Service and Component Capacity

This activity manages the performance and capacity of Services, components, and resources by monitoring, analyzing and tuning to make the most efficient use of existing IT resources. It uses the outputs from CapM2 and CapM3. Analysis of the monitored data using trending, forecasting, modeling techniques, and thresholds may identify areas of configuration that can be tuned for improved Service, system, and component resource utilization or the performance of a Service.

This activity also includes the first step to determine if any new Service, upgrade or demand can be met with current resources before going to a material solution in CapM5. The outputs from this activity are used in CapM6 as part of the Capacity Plan. Sub-processes of this activity are also where overages in capacity are handled.

[CapM5] Plan New Capacity

This activity is a continuous, iterative process that produces a Capacity Plan to document current levels of resource utilization and Service performance. Analysis and data from CapM4 help determine what procurements need to be made to satisfy requirements. This activity involves the use of trending, forecasting, modeling techniques, and thresholds to plan upgrades, enhancements, and estimated future requirements. It becomes a tool that reflects Capacity Management goals by incorporating the current business operation and requirements. The plan should be updated to forecast future requirements for resources that support all Services (existing and new) that are based on business requirements.

[CapM6] Monitor, Manage and Report Capacity Management

This activity supports continuous monitoring and analysis of operational results data and comparison with Service achievement reporting to identify Capacity Management trends and issues. Capacity Management information is used to generate detailed Service component reporting and provide perspective on overall CapM process performance.

[CapM7] Evaluate Capacity Management Performance

This activity describes the tasks required to assess the efficiency and effectiveness of the Capacity Management process. It includes the capture of information on records, the relationship with other process areas, and the suitability of procedures and training. It is used as a basis to ensure that the Capacity Management process remains fit for purpose and identifies where changes to the process might be required.

5). Information Security Management (ISM)

Purpose - The purpose of the Information Security Management (ISM) process is to manage information security at an approved level of security within all Service Management activities. This includes compliance with the organization-specific information security requirements. ISM ensures that security controls required to perform Service Management activities effectively protect information assets. This includes preserving the Confidentiality, Integrity, and Availability (CIA) of all data transported.

- **Confidentiality** represents how the data and information must be accessible only to its predefined recipients
- **Integrity** represents how the data and information that must be correct and complete
- **Availability** represents how the data and information must be accessible as needed

Scope - The scope of ISM includes all use and misuse of all IT systems that support the business and Services. This is done from four aspects:

- **Personal** – Defines security policies as related to human resources and staff awareness and responsibilities
- **Procedural** – Procedures to control security that flow from the security policy and process
- **Physical/Facilities** – Controls used to protect any physical sites against security Incidents
- **Technical** – Controls used to protect the IT infrastructure against security Incidents

Process Benefits

- Information security awareness is heightened
- Data is protected, accurate, and available when needed
- Effective access to information by authorized personnel
- Identification of potential security vulnerabilities before they can cause a security-related
- Incident
- Information exchanges can be trusted
- With Incident Management, Incidents are detected and managed in a controlled manner
-

Expected Outcomes

- Information security requirements are identified and established
- Information security risks are identified and assessed
- Assessment criteria for Information Security risks and risk appetite are identified
- Information security risks measures are defined and applied
- Information security Incidents are identified and recorded

- Information security concerns are communicated to stakeholders and interested parties
- The impact of changes on Information Security is evaluated and reported

Process Activities

[ISM1] Establish Information Security Management Framework

Implementation of this process varies widely across industries and organizations. Some organizations keep the process solely related to 'information assets' and even go further to more narrowly scope what constitutes the information that this process addresses. Process scopes can include or exclude aspects of Cyber Security as well or define a specific Cyber Security process. Other organizations expand this process to not only address cyber-security aspects but also to include a broader scope of security including personnel security, physical security, operations security, industrial security etc. This activity defines all direction, guidance, policies, and procedures for how the process will be performed. All of this is collectively referred to as the "ISM process framework" and is used as reference information for all other activities. This information is reviewed in the Evaluate Process Performance activity, which generates recommendations for changing and improving the ISM process framework.

[ISM2] Create and Sustain Security Policy or Directive

This activity incorporates the goals and objectives for security that needs to be established. It maintains relevancy as circumstances change for the Service Provider and its customer set.

[ISM3] Categorize for Certification and Accreditation

This activity assigns or verifies the security classification level of information assets to support Certification and Accreditation.

[ISM4] Analyze Security Threats, Vulnerabilities, and Risks

This activity identifies security threats, vulnerabilities, and risks. It includes mitigation recommendations based on analysis and policy guidance from applicable security instructions.

[ISM5] Plan and Implement Security Practices

This activity establishes the Security Plan in compliance with applicable security instructions. It defines and creates an appropriate security infrastructure and procedures, translates actions in the plan to security directives, and communicates them to the appropriate audiences.

[ISM6] Direct/Perform Security Protection Operations

This activity executes prescribed information security controls and procedures by operating and activating protections within IT solutions and Services. It monitors the full range of information security measures and capabilities, responds to Service or authorization requests, and monitors real-time intrusion prevention or detection with established response criteria. Additionally, this activity notes information security violations and initiates incidents when required.

[ISM7] Monitor, Manage and Report Information Security Management

This activity addresses a regular review of security controls and mechanisms and determines whether they effectively implement security policies and procedures as described in applicable security instructions. This activity works hand-and-hand with [ISM 6], as it manages the documented information security violations. Security assessments, inspections, and audits occur in this activity.

[ISM8] Evaluate Information Security Management Performance

This activity describes the tasks required to assess the efficiency and effectiveness of the ISM process. It includes the capture of information on records, the relationship with other process areas, and the suitability of procedures and training. It is used as a basis to ensure that the process remains fit for purpose and identifies where changes to the process might be required.

The Information Security Management System (ISMS)

Each of the following requirements should be addressed within the ISMS:

- ISMS Established
    - Shall have defined the scope and boundaries of their specific ISMS in terms of the characteristics of the business, location, assets and technology, and a method to address exceptions.

- o The ISMS policy is defined in terms of the characteristics of the business, its organization, location, assets, and terminology, which includes a framework for setting objectives, considers business and legal or regulatory requirements, and contractual security obligations.
- Risk Assessment Approach Defined:
  - o The Risk Assessment approach for organization and provider shall include Risk Assessment methodology specific to the ISMS, business information security, legal and regulatory requirements, as well as acceptance criteria for accepting risk and levels of risk. The Risk Assessment methodology selected shall ensure that Risk Assessments produce comparable and reproducible results.
- Risk Identification:
  - o The assets and asset owners are identified within the scope of the ISMS.
  - o The threats, vulnerabilities and the impact that losses of Confidentiality, Integrity, and Availability may have on the assets have been defined.

- Analyze and evaluate the risks:
  - The business impacts upon the organization that may result from security failures have been assessed. The business impact considers the consequences of a loss of Confidentiality, Integrity or Availability of the assets.
  - The levels of risk have been estimated.
  - There is a determination whether the risks are acceptable or require treatment using the criteria established.
- Identify and evaluate options for the treatment of risks (Actions Include):
  - Applying appropriate controls
  - Accepting risks, providing they satisfy the organization's policies and criteria for accepting risks
  - Avoiding Risks
  - Transferring the associated business risks to other parties, e.g. insurers, suppliers.
  - Control objectives and controls for the treatment of risks are selected

- Control objectives and controls shall be selected and implemented to meet the requirements identified by the risk assessment and risk treatment process.
    - Leadership approval has been obtained for the proposed residual risks.
    - Leadership authorization to implement and operate the ISMS is obtained.
- A Statement of Applicability shall be prepared that includes the following:
    - The control objectives and controls currently implemented
    - The exclusion of any control objectives and controls and justification for their exclusion.

6). IT Service Continuity Management (ISM)

Purpose - The purpose of the IT Service Continuity Management process is to manage the risks that could affect critical Services and ensure that there is a plan to recover minimum agreed business continuity-related Service Levels.

Scope - ITSCM is responsible for safeguarding the interests of all stakeholders served. It identifies risks, minimizes the impact of Service disruptions and ensures that the required technical and Service facilities can be recovered within required and agreed timeframes. It includes plans to provide agreed upon Service Levels in exceptional circumstances. ITSCM is proactive in supporting the plan to avoid disaster situations and reactive in executing the plan in response to major events. Periodic testing of the ITSCM should be conducted.

Process Benefits

- Controlled recovery of systems
- Better Risk Assessments
- Better understand and address weaknesses that may affect the business stakeholders before a disaster occurs
- Better Service impact analysis
- Identification of critical Services and functions

- Prioritization of Services allow for better utilization of resources during recovery
- Tested plans reduce downtime in the event of a disaster
- Confidence that needs of the business can be fulfilled under less than ideal conditions
- The IT Service Continuity Plan sets procedures that are regularly tested and updated to prevent, address, and recover from major disruptions and loss of critical Services for extended periods
- Reduction in overall risk of failures in the production environment
- Business partner confidence in ability to provide support in a crisis
- Support of Business Continuity Management

Expected Outcomes

- Service continuity requirements are identified
- A Service Continuity Plan is developed using the Service Continuity requirements
- Service continuity is tested against the Service Continuity requirements to validate the plan
- Changes to Service Continuity requirements are reflected in the Service Continuity Plan

Process Activities

[ITSCM1] Establish IT Service Continuity Management Framework

This activity defines all direction, guidance, policies, and procedures for how the process will be performed. All of this is collectively referred to as the "ITSCM framework" and is used as reference information for all other activities. This information is reviewed in the Evaluate Process Performance activity, which generates recommendations for making changes and improvements to the ITSCM framework.

[ITSCM2] Identify Continuity Requirements

This activity identifies those requirements that are critical to continuing operations at the level required for the essential functions of the business customer. The activity continues with a Risk Assessment that identifies what might occur in the event of a disruption or degradation.

[ITSCM3] Create and Maintain IT Service Continuity Strategy

This activity is responsible for identifying risk reduction measures for the identified continuity requirements and establishing what countermeasures and recovery options exist to support these requirements. It considers the types of risk that might be encountered, and the potential costs involved for each recovery option. The outcome of this activity is an agreed to IT Service Continuity Strategy and a set of IT Service Continuity requirements.

[ITSCM4] Create and Maintain IT Service Continuity Plan (ITSCP)

This process is responsible for identifying the resources (e.g. people, processes, technology, facilities, and communications) necessary to support the required Services if the ITSCP is invoked. This activity also identifies the actions necessary for successful invocation of the plan. It is responsible for the ongoing maintenance of the plan and considers changes to essential functions and changes to the infrastructure.

[ITSCM5] Prepare IT Service Continuity Capability

This process ensures that an invocation of the ITSMP results in the ability to recover and restore required Services to a predetermined level, and in a predetermined timeframe. It has the responsibility for ensuring that all plans are tested regularly, both on a planned and unplanned basis; that the process passes audit requirements, and that the results from tests are captured and fed back to other processes to ensure that the ITSCP remains fit for purpose.

[ITSCM6] Execute IT Service Continuity Plan (ITSCP)

This process is responsible for implementing the ITSCP according to predetermined criteria. It is responsible for maintaining business operational requirements for an unspecified amount of time, and for ensuring a controlled restoration to normal service.

[ITSCM7] Monitor, Manage, and Report IT Service Continuity Management

In this activity, ITSCM activities are monitored to determine whether suitable progress is being made. Results are reported, and unsatisfactory results may lead to a review of ITSCM actions. In addition, responses are provided to requests for information and status of the ITSCM process.

[ITSCM8] Evaluate IT Service Continuity Management Performance

This activity describes the tasks required to assess the efficiency and effectiveness of ITSCM. It includes the capture of information on records, the relationship with other process areas, and the suitability of procedures and training. It is used as a basis to ensure that the ITSCM process remains fit for purpose and identifies where changes to the process might be required.

7). **Service Level Management (SLM)**

Purpose - The purpose of Service Level Management (SLM) is to provide a framework for regular contact between the business customer and the Service Provider to negotiate and document Service Level targets and responsibilities. Service Level Agreements (SLAs) and Operational Level Agreements (OLAs) are developed to understand specific and measurable targets regarding the level of Service quality. SLAs are supported by OLAs and Underpinning Contracts (UCs).

Scope - The scope of SLM is a reciprocal relationship and representation of the IT Service Provider to the business customer and the business customer to the IT Service Provider with regards to Service quality. A clear and unambiguous expectation to the level of Service being delivered is paramount to ensure business customer satisfaction. The process coordinates the amount and availability of Service components for an entire Service to enable delivery of the Service requirements and agreed Service Level Objectives to the stakeholder. It monitors and reports on the Service Levels attained.

Process Benefits

- The culture will establish a Service-value, Service-oriented viewpoint
- Financial savings through improved Service quality and better resource usage in resolving outages
- Service Provider and business customer will better understand each other's responsibilities related to Services
- Providers and business customers develop mutually beneficial relationships and deliver relevant Services that improve business partner satisfaction
- Improved planning based on user agreements
- Improved management by focusing on Service delivery and business goals
- Services are continually and consistently monitored and measured quantitatively and qualitatively

Expected Outcomes

- Services and dependencies are identified
- Service Level Objectives and workload characteristics for Services are defined in SLAs
- Services are monitored against SLAs
- Service level performance against Service Level Objectives is communicated to interested parties
- Changes to Service requirements are reflected in the SLAs

Process Activities

[SLM1] Establish Service Level Management Framework

This activity defines all direction, guidance, policies, and procedures for how this process will be performed. All of this is collectively referred to as the "SLM process framework" and is used as reference information for all other activities. This information is reviewed in the Evaluate SLM Performance activity, which generates recommendations for changes and improvements to the SLM process framework.

[SLM2] Capture Service Level Requirements

This activity facilitates the discussions with business partner stakeholders to capture desired Service Level Requirements and Service Level targets. These requirements are reflected in the various agreements used to support the Service, such as Service Level Agreements (SLAs), Operation Level Agreements (OLAs), and Underpinning Contracts (UCs).

[SLM3] Review Existing OLAs and UCs

This activity reviews the outlines of the required OLAs and UCs required to support a new service to determine if OLAs or UCs already exist that will meet the technical requirements.

[SLM4] Define Requirements for OLA or UC

This activity defines the complete requirements for new OLAs and/or UCs or modifications to existing OLAs and/or UCs. Technical requirements must have clearly defined boundaries and handoffs.

[SLM5] Negotiate or Archive SLA

Formalized requirements are negotiated between the Service Provider and business customer of the Service into new or modified SLAs. In addition, SLAs that are no longer needed are archived. New and modified SLAs and UCs are published to the appropriate repositories and associated with corresponding Services.

[SLM6] Monitor and Report Service Level Achievements

This activity is the continuous monitoring of Service Level achievements. The data is collected from various systems and tools. SLA data information (from Service Providers, monitoring applications, and stakeholder feedback) is run through reporting mechanisms to determine if SLA targets were met or missed.

[SLM7] Conduct Service Review

Using Service Level Achievement Reports, an analysis of the SLAs/OLAs/UCs is conducted to reveal and assess existing and potential gaps between target and actual Service delivery or Service Level achievements. Any penalties are identified during these reviews.

[SLM8] Formulate Service Improvement Plan

A Service Improvement Plan (SIP) is created from results of the Service Level achievement review, stakeholder feedback, and Service delivery units, regarding improvement suggestions. The SIP focuses on recommendations for SLA compliance improvements and specific target modifications.

[SLM9] Monitor, Manage and Report Service Level Management

This activity supports continuous monitoring and analysis of operational results data and comparison with Service achievement reporting to identify Service Level Management trends and issues. Service Level Management information is used to generate detailed Service component reporting as well as a perspective on overall Service Availability.

[SLM10] Evaluate Service Level Management Performance

This activity describes the tasks required to assess the efficiency and effectiveness of the Service Level Management process. It includes the capture of information on records, the relationship with other process areas, and the suitability of procedures and training. It is used as a basis to ensure that the Service Level Management process remains fit for purpose and identifies where changes to the process might be required.

8). Supplier Management (SUP)

Purpose - The purpose of the Supplier Management process is to ensure that supplier Services are integrated into Service delivery to meet the approved requirements. It ensures that suppliers are managed to support the business and Service Level targets. Objectives include:

- Obtain maximum value for the money spent on suppliers
- Ensure that contracts are aligned with IT and organizational strategy and support the various aspects of Service Level Management

Scope - Suppliers are horizontally or vertically integrated participants in the supply chain of a Service. Therefore, the process ensures that the Service Provider establishes commitments with suppliers who support the integration and alignment of Services and agreements between the Service Provider and stakeholders. It ensures that the resources provided by the supplier adequately fulfill the IT Service requirements as defined by the SLM process and verifies that suppliers can demonstrate management of subcontracted partners to meet obligations and contractual requirements. The scope includes:

- Implementation and enforcement of a supplier policy
- Supplier and contract categorization and risk assessment
- Supplier and contract evaluation and selection
- Development, negotiation, and agreement of contracts
- Contract review, renewal, and termination
- Management of suppliers, supplier performance, and contractual dispute resolution
- Agreement and implementation of Service and supplier improvement plans
- Maintenance of standard contracts, terms, and conditions

Process Benefits

- Ensures that underpinning contracts and agreements with suppliers support and align with business needs, Service Level Requirements (SLRs), and Service Level Agreements (SLAs)
- Obtain maximum value for supplier Services
- Creation and management of supplier and contract information

Expected Outcomes

- Relationships between the Service Provider and suppliers are managed
- Services to be provided are negotiated with each supplier
- Roles and relationships between suppliers is determined
- Supplier obligations to meet Service requirements are monitored
- Supplier performance against approved criteria is monitored
- The capability of subcontracted suppliers to meet obligations is confirmed

Process Activities

[SUP1] Establish Supplier Management Framework

This activity defines all direction, guidance, policies, and procedures for how the process will be performed. All of this is collectively referred to as the "SUP process framework" and is used as reference information for all other activities. This information is reviewed in the Evaluate Process Performance activity, which generates recommendations for making changes and improvements to the SUP process framework.

[SUP2] Define Business Case and Invite Proposals

Define initial business case; includes costs, timelines, targets, value, and risks. Invite suppliers to provide proposals and/or bids for meeting defined business needs. Ensure that draft proposals conform to strategy and policy.

[SUP3] Evaluate Potential Suppliers and Award Contract

Evaluate potential suppliers, identify alternatives, and select suppliers. Negotiate terms and conditions, responsibilities, resolution of disputes, renewals and extensions, and other contract content. Award selected supplier the contract.

[SUP4] Establish New Supplier and Contract

Integrate new suppliers by providing supplier-appropriate access to necessary systems and data. Initiate supplier contracts and relationships.

[SUP5] Provide Supplier and Service Information

This activity provides information about supply items, such as a supply item catalog (hardware, software, services, and external resources that contains information about supply items,) potential suppliers for those items (including supplier priorities and options) and supply item availability.

[SUP6] Manage Supplier Delivery

This activity manages supplier delivery and evaluates supplier performance. During which, the review of supplier delivery against business, technical and financial criteria is performed. Relationships during delivery periods, including communication, risks, changes, failures, improvements, contracts, and interfaces are maintained. During this activity, supplier performance is periodically reviewed and assessed against business needs, targets, and agreements. The recommendation of possible delivery closure, renewal, or extension is given, as applicable.

[SUP7] Renew or Terminate Contract

In this activity, negotiations of the renewal, termination, or transfer of contracts with the supplier are conducted. If a contract is terminated or transferred, this activity manages the completion of the supplier relationship.

[SUP8] Monitor, Manage and Report Supplier Management

In this activity, all process activities are monitored to determine whether suitable progress is being made. Unsatisfactory results are reported and may result in intervention into process work. In addition, responses to requests for information and status about the process are provided.

[SUP9] Evaluate Supplier Management Performance

This activity describes the tasks required to assess the efficiency and effectiveness of the Supplier Management process. It includes the capture of information on records, the relationship with other process areas, and the suitability of procedures and training. It is used as a basis to ensure that the Supplier Management process remains fit for purpose and identifies where changes to the process might be required.

Chapter 8 – Service Transition

The Service Transition phase of the Service Lifecycle is responsible for assisting in the design of new or changed Services as they move from concept to production, the implementation or decommissioning of Services or Service components, and making modifications to Services as a result of required corrective actions, or to improve an existing Service. As such, it is the responsibility in this Lifecycle phase to ensure that the strategic vision of the organization is carried out and includes ensuring that the creation of Services in Service Design is carried out during the implementation phases.

1). **Transition Planning and Support (TPS)**

Purpose - The purpose of Transition Planning and Support is to plan and coordinate the resources to take a new or changed Service, or a Service to be decommissioned (decided in Service Portfolio Management process) through Release and Deployment into the production environment, ensuring that the effort is accomplished in accordance with predicted cost, quality, time estimates, and acceptable levels of risk, and that it meets all requirements in the Service Design Package.

Scope - TPS ensures that the Service components are effectively integrated into a new or changed Service and the Service Provider and business customer are prepared to operate the solution to deliver the desired outcomes.

Process Benefits

- Ensures integrity of business customer and related Service Assets
- Coordinated activities across projects, suppliers and Service teams
- Single point of communication related to Service activities in scope
- Reduction in variation from requirements to production
- Ability to deliver higher volumes of change at higher success rates
- Reduced variation in release schedule adherence due to standardized, holistic planning
- Improved integration of Services with the business customer's needs
- Consolidated deployment process
- Better planning and resource allocation
- Improved risk management, resulting in reduced adverse impact due to increased predictability of quality of Service
- Better integration of supporting processes

Expected Outcomes
- Requirements for Service Transition are identified and approved
- New or changed methods, procedures, and measures for the new and changed Service(s) are identified
- New or changed knowledge, skills, and abilities are identified, approved, acquired and assigned
- Transition activities to be performed by Service Provider or client are identified, approved & executed
- New or changed plans for Availability, IT Service Continuity, Capacity and Information Security are identified, communicated and employed (these are also identified with Service Design Coordination)
- New or changed authority and responsibility for the new and changed Services are identified
- New or changed contracts and formal agreements with internal groups and suppliers to align with the updated requirements are identified and employed
- Resources for the delivery of the new or changed Services are identified and provided
- The new or changed Service is deployed and tested according to relevant Service specification
- The new or changed Service is accepted in accordance with established Service acceptance criteria

- Communicate information regarding the outcomes of the transitioned Service to interested parties

Process Activities

[TPS1] Establish Transition Planning and Support Framework

This activity defines all direction, guidance, policies, and procedures for how the process will be performed. All of this is collectively referred to as the "TPS process framework" and is used as reference information for all other activities. This information is reviewed in the Evaluate Process Performance activity, which generates recommendations for changes and improvements to the TPS process framework.

[TPS2] Define Service Transition Plan

As a blueprint for how the transition is carried out, the transition plan describes the activities needed to carry out the transition, as well as resource modifications, schedules, organizational changes, training, risks, communications, and other important considerations. The transition plan is used throughout the new or changed Service transition.

[TPS3] Initiate Transition Change Requests

In this activity, all Requests for Change (RFC) needed for the Service transition are created and submitted to the Change Management process. The RFCs are created with the appropriate sequencing and timing to properly design the transition.

[TPS4] Guide Service Transition

As the transition-related Request for Change is executed, this activity provides support for deployments and other implementations related to the Service transition. This includes ensuring that acquisitions related to the transition are completed on-time, Release deployments are sequenced and coordinated properly, communications related to the transition are performed, pilots (if required) are carried out, post-installation testing occurs, and other transition-related tasks are performed.

[TPS5] Adjust Resources and Train

Resources are added or removed as needed for the transition of the Service. These resources include operations and support personnel. Early Life Support (ELS) may be considered in resource adjustment. Users and other Service-related personnel are provided job-appropriate training.

[TPS6] Review and Close Service Transition

The results of the Service transition are reviewed to determine if the transition was carried out as intended. Deviations and deficiencies in the transition are addressed, possibly resulting in additional RFCs. When transition issues have been adequately addressed, the Service transition is closed.

[TPS7] Monitor, Manage and Report Transition Planning and Support

This activity supports continuous monitoring and analysis of operational results data and comparison with Service achievement reporting to identify Transition Planning and Support trends and issues. Transition Planning and Support information are used to generate detailed Service component reporting as well as a perspective on overall Service availability.

[TPS8] Evaluate Transition Planning and Support Performance

This activity describes the tasks required to assess the efficiency and effectiveness of the Transition Planning and Support process. It includes the capture of information, the relationship with other process areas, and the suitability of procedures and training. It is used as a basis to ensure that the Transition Planning and Support process remains fit for purpose and identifies where changes to the process might be required.

2). **Asset Management (AM)**

Purpose - The purpose of Asset Management is to manage the finances, contracts, and usage of IT assets throughout their lifecycles to balance Service requirements, total costs, budgeting, and compliance. The lifecycle ranges from procurement through deployment to use (and upgrades) to decommissioning (or reuse) to disposal. The difference between Configuration Management and Asset Management is that Configuration Management is concerned with the relationships between Configuration Items (CIs) in support of the Services while Asset Management manages the financial attributes of the asset such as costs, compliance, etc.

Asset Management may also manage the assets of organizations not directly related to IT support of a Service. In some instances, the Asset and Configuration Management processes are one process, not two separate processes. And in some cases, the Asset Management database becomes part of the Configuration Management System. The needs of the organization drive the decision with regards to having one or two processes. The relationship of assets to Services is covered under Configuration Management.

Scope- The scope is unique to the organization based on the established purpose of the process and needs of the organization. It can be confined to assets that directly affect Services provided or can be as broad as to include physical assets as well. Thus, Asset Management manages or could be responsible for managing:

- Hardware (including maintenance)
- Software (including maintenance)
- Software Licenses (issuance accountability)
- Facilities (and related, such as desks, etc.)

Process Benefits

- Creates improved procurement processes through centralization of all asset data and asset-related financial information
- Simplifies inventory and auditing processes
- More accurate risk assessments due to better asset tracking
- Improved understanding of the real cost of assets
- Improved adherence to vendor licensed software products
- Increased insight into the Total Cost of Ownership of IT Services through detailed asset information

Expected Outcomes

- Asset information is available on which to base business decisions
- Existing investments in hardware, software, and licenses are used
- Ensured compliance with statutes, regulations, directives and enterprise architecture
- Audit and governance compliance conformance is assured

- Full control of all assets is assured throughout the asset lifecycle
- Reduction in unnecessary or duplicate expenditures
- Assets are available at the right time for deployment
- Total Cost of Ownership and Return on Investment can be calculated

Process Activities

[AM1] Establish Asset Management Framework

This activity defines all direction, guidance, policies, and procedures for how the process will be performed. All of this is collectively referred to as the "AM process framework" and is used as reference information for all other activities. This information is reviewed in the Evaluate Process Performance activity, which generates recommendations for making changes and improvements to the AM process framework.

[AM2] Record and Control Assets

This activity prepares assets for use and includes receipt of assets from the supplier or when repurposing or redeploying existing assets. The activity also provides the status of assets and pre-deployment actions, such as imaging and asset identification tags, assignment of assets and when applicable, transportation coordination of assets to new locations. This activity also executes the retirement and disposal of assets.

[AM3] Maintain Asset Record Information

The purpose of this activity is to maintain asset records: change, update, or delete asset data as required. Incident Management, Problem Management, and Configuration Management can trigger modifications to asset data. This activity also administers the asset database and performs asset reconciliation. The asset database includes all assets with a status designation such as ordered, in storage, assigned, retired, or disposed of, etc.

[AM4] Monitor, Audit and Reconcile Asset Records

In this activity, the status of IT assets is monitored. Compliance status for licensing and information security requirements is also monitored. Formal inventory audits of all physical assets occur in this activity. Additionally, audits of the Asset Management System and audit reconciliation are performed. Audits of logical assets include installed software on workstations and IT configurations or as required by the organization.

[AM5] Conduct Asset Remediation

This activity performs reporting and oversight for all assets requiring remediation, including remediation activities for missing and deployed assets. The goal of this activity is to ensure that assets which cannot be physically verified are accurately reflected in the Asset Management System. Assets may be marked as active, retired, missing, or deployed.

[AM6] Retire and Dispose of Assets

This activity ensures that all assets meet criteria for retirement and are returned to storage in preparation for disposal. Assets that have reached end-of-life are disposed of as required. Asset records and databases are updated with new status information.

[AM7] Monitor, Manage and Report Asset Management

In this activity, Asset Management activities are monitored to determine whether suitable progress is being made. Results are reported, and unsatisfactory results may lead to a review of Asset Management actions. In addition, responses are provided to requests for information and status of the Asset Management process.

[AM8] Evaluate Asset Management Performance

This activity describes the tasks required to assess the efficiency and effectiveness of the Asset Management process. It includes the capture of information, the relationship with other process areas, and the suitability of procedures and training. It is used as a basis to ensure that the Asset Management process remains fit for purpose and identifies where changes to the process might be required.

3). Change Management (ChM)

Purpose - The purpose of the Change Management (ChM) process is to ensure that all Changes are assessed, approved, implemented and reviewed in a controlled manner. To this end, Change Management ensures that any change to the IT production environment, whether it involves an addition, modification, or deletion of a Service or Service component, is in line with the overall organization strategy. This process provides standardized methods and procedures for the efficient and prompt handling of technical changes to minimize the impact of change-related Incidents to Service quality and improves day-to-day operations of the organization.

Scope - The scope of ChM encompasses any asset or Configuration Item (CI) that supports a service. Thus, ChM is responsible for managing the Change process involving hardware (infrastructure), software and all documentation associated with running, supporting and maintaining production systems. All changes are planned and controlled to ensure timely updates with no unnecessary disruption or unintended consequences.

Process Benefits

- Consistent tracking, scheduling, and documentation of the addition, modification or retirement of CIs
- As the Change moves through its lifecycle, it's status is visible
- Early identification of risk: The process includes submission of a risk analysis with every major Change. This proactive approach mitigates risks to cause the least impact to business customer's Service.
- Improved prioritizing and response to business and Service Provider Change proposals
- Implemented Changes that meet business customer-agreed Service requirements with optimized costs
- Contributes to governance, legal, contractual and regulatory requirements
- Fewer failed changes and therefore reduction in Service disruption, defects, and re-work
- Provides change history throughout the Service Lifecycle
- Aids productivity of staff through minimizing disruptions due to high levels of unplanned or "Emergency" Changes and hence maximizes Service Availability

Implementing and Improving ITSM

- Reduces the Mean Time to Restore Service (MTRS), via quicker and more successful implementations of corrective changes
- Reduces risks associated with introducing Change to the environment
- Reduces unplanned work due to reduction in Incidents caused by Change
- Increase the identification and approval of Standard Changes, allowing for more efficient and timely implementations

Expected Outcomes

- Requests for Change are recorded and categorized
- Requests for Change are assessed using defined criteria
- Requests for Change are approved before resources are committed to develop and deploy the Change
- A schedule of Changes and Releases is established, maintained and communicated to interested parties
- Approved Changes are developed and tested
- Unsuccessful Changes are reversed or remedied

Process Activities

[ChM1] Establish Change Management Framework

This activity defines all direction, guidance, policies, and procedures for how the process will be performed. All of this is collectively referred to as the "Change Management process framework" and is used as reference information for all other activities. This information is reviewed in the Evaluate Process Performance activity, which generates recommendations for changes and improvements to the Change Management process framework.

[ChM2] Create and Record Change Request

This activity involves formulating and storing the information about any change. Each Change Request will be accompanied by a defined outline of information established for assessment and other Change Management activities. Information can vary depending on the context, scale, and potential impact of the requested Change.

[ChM3] Accept and Categorize Change

This activity examines the Request for Change (RFC) to determine if it should be accepted for consideration. RFC acceptance requires all information to be logged. Incomplete information can cause an RFC to be returned for additional or amplifying information. After initial acceptance, the RFC is categorized.

[ChM4] Evaluate Change

Each Change is analyzed to determine the impact on existing and planned CIs and the impact on resources required to build and deploy the Change. This involves identifying the appropriate Change model for handling the Change, verifying appropriate Change authority when necessary, scheduling a Change Advisory Board (CAB) meeting if specified by the Change model, and obtaining a complete set of analysis results and issues. Assessment often assigns impact categorization classes such as minor or major.

[ChM5] Authorize and Schedule Change

This activity represents a decision checkpoint against the Change based on impact. It examines the analysis results from the Evaluate Change activity and determines whether the Change should be approved. If approved, the Change deployment approach and targeted Change deployment schedule are determined for the Change. The way the Change is approved will depend on the organization structure, but formal approval will be obtained for each Change from the Change authority (Change Manager or applicable Change Advisory Board). The activity for scheduling a Change considers the Change Schedule, eliminating conflict between differing Changes, and assigning appropriate resources accordingly.

[ChM6] Coordinate Change Implementation

This activity coordinates implementation of the change. If the approved Change created or updated a solution the solution components must first be built and tested. Approved Changes are made available primarily through Release and Deployment Management (RDM); however, some Changes are implemented through assignment by the Change Manager (within Change Management). This determination is made by Change Management policies and the appropriate change model. Change Management monitors the deployment of the Change, as carried out by RDM.

[ChM7] Evaluate and Close Change

This activity contains the tasks involved in reviewing all implemented Changes (including Post-Implementation Review) after a predefined period has elapsed or another review trigger has been activated. It ensures that the Change exhibits the desired effect and meets objectives, and that users and customers are satisfied with the results or identifies any deficiencies. The review activity determines whether the implementation plan and the back-out plan, as appropriate, are performed correctly, and whether the Change was implemented on time and to cost. It determines whether any follow-up action (such as the creation of a new Request for Change) is required. Subsequently, a formal close of the Change is performed. The closure of a Change includes updating other processes with the change status.

[ChM8] Monitor, Manage and Report Change Management

Continuous monitoring and analysis of operational results and comparison with Service achievement reporting identifies Change Management trends and issues. Change Management data is used to generate detailed Service component reporting as well as a perspective on Service Availability.

[ChM9] Evaluate Change Management Performance

This activity describes the tasks required to assess the efficiency and effectiveness of the Change Management process. It includes the capture of information, the relationship with other process areas, and the suitability of procedures and training. It is used to ensure that the Change Management process remains fit for purpose and identifies where changes to the process might be required.

4). Change Evaluation (EVAL)

Purpose - Change Evaluation is a formal evaluation process that is conducted prior to the execution of any Changed the organization deems significant. The organization determines the definition (threshold) of significant Changes that invoke this process. The goal of Change Evaluation is to provide accurate information to the Change Management process as to the impact and effect the Change may have on Service capability prior to acceptance of the Change.

Scope - The scope of the Changes to be formally evaluated is determined by the organization. As a guideline, this can include any Change that introduces a new Service, causes a substantial Change to an existing Service, or retires a Service.

It may also be determined by impact, or by a project that impacts support, such as a reorganization or Service Desk consolidation. Resources, in time, equipment or money, may also be a consideration in determining if this process should be invoked from the Change Management process. When the Change Evaluation process ends, the Change Management process takes responsibility for further change activities.

Process Benefits

- Additional focus and governance of significant Changes
- Proper command and control of major Changes
- Multiple risk analysis with each significant Change
- Better allocation of resources
- Significant Changes may undergo multiple risk analysis as they move through the Change lifecycle
- Transparency into the status of the Change

Expected Outcomes

- All factors are considered prior to making a major Change, including capability, tolerance for risk, organizational structure, resources, modeling, people, and all other projects and Changes
- Major Changes are viewed through Service filters, not simply as IT projects

This process is invoked as a part of the Change Management process at the discretion of the organization. Thresholds are determined by the organization as to when this process is needed and executed.

5). Configuration Management (CfM)

Purpose - The purpose of Configuration Management (CfM) is to control, identify, record, and report IT components, including versions (where appropriate), constituent components, states and most importantly, relationships to other IT components and Services.

Scope - Configuration Items (CIs) are any assets that need to be managed to deliver a Service. CIs that should be under the control of Configuration Management include hardware, software, systems, Services, applications, their relationships, and associated or related documentation, (e.g., Service Level Agreements). Configuration Management establishes and maintains the integrity of Services and their configuration information to enable effective control of the Services and to reduce the risk of unintended consequences during change execution.

Process Benefits

- Accurate information on CIs and their documentation: This information supports all other IT Service Management processes, such as Release Management, Change Management, Incident Management, Problem Management, Capacity Management, and any necessary contingency planning. Configuration Management can provide information for upgrade planning and replacements.
- Facilitates adherence to legal obligations: Configuration Management maintains an inventory of all software and hardware within an IT infrastructure.

- Improves security by controlling versions of CIs in use: This makes it more difficult for those CIs to be changed accidentally, maliciously, or for erroneous versions to be added.
- Allows the organization to perform impact analysis and schedule changes safely, efficiently, and effectively: This reduces the risk of Changes that may negatively impact the live environment.
- Unified view into the relationships between CIs, which correlates to the following processes as well: Asset, Configuration, Change, Event, Problem, and Incident Management
- Better risk assessment for approving Changes
- Better Incident Management, since failing components are traceable to Services

Expected Outcomes

- All configuration items (CIs) within IT systems and infrastructure are accurately identified and relationships recorded
- The status of the CIs and modifications are effectively recorded, tracked, and reported
- Changes to CIs are controlled

- Any exceptions between configuration records and the corresponding CIs are identified and corrected
- The integrity of released systems, Services and Service components is assured
- The configuration of released systems, Services and Service components is controlled

Process Activities

[CfM1] Establish Configuration Management Framework

This activity defines all direction, guidance, policies, and procedures for how the process will be performed. All of this is collectively referred to as the "CfM process framework" and is used as reference information for all other activities. This information is reviewed in the Evaluate Process Performance activity, which generates recommendations for changes and improvements to the CfM process framework.

[CfM2] Perform Configuration Identification

This activity identifies, defines and records the types of CIs under the control of Configuration Management, the CI naming conventions, attributes, relationships to other CI types, data integrity rules, and requirements and design documentation.

[CfM3] Conduct Configuration Control

This activity ensures that CIs and relationships and status are recorded accurately throughout each CI lifecycle. It generates configuration baselines and manages drift within acceptable limits. A baseline must be created to help restore a set of CIs to a known stable state if a change fails and its back-out plan is implemented.

[CfM4] Report Configuration Status

This activity makes CI information available to authorized requestors. The information ranges from detailed CI attributes and relationships to summarized information. It may cover an individual CI or a collection of CIs. CI information is provided in line with a planned schedule or in response to an individual request.

[CfM5] Conduct Configuration Verification & Audit

This activity ensures that CI information matches the physical reconciliation data, that naming conventions are adhered to and that the Definitive Media Library (DML) and/or secure repositories agree with the CI information. The audit is performed regularly, as stipulated by the Configuration Management Plan, or as requested by the Configuration Manager or other authorized personnel.

[CfM6] Monitor, Manage and Report Configuration Management

In this activity, all Configuration Management activity is monitored to determine whether suitable progress is being made. Unsatisfactory results are reported and may result in actions taken to address any issues.

[CfM7] Evaluate Configuration Management Performance

This activity describes the tasks required to assess the efficiency and effectiveness of the Configuration Management process. It includes the capture of information, the relationship with other process areas, and the suitability of procedures and training. It is used as a basis to ensure that the Configuration Management process remains fit for purpose and identifies where changes to the process might be required.

6). **Knowledge Management (KM)**

Purpose - The purpose of the Knowledge Management is to ensure that the right information is delivered to the appropriate place or person at the right time to enable informed decisions that improve performance, make the enterprise more efficient and to better serve the IT department and business customer.

Knowledge Management provides the mechanism to help create, capture, share and act upon information in ways that will measurably improve the delivery and support of Services as defined by IT leadership vision. KM is the instrument to improve the IT department's ability to execute core competencies in support of the business.

Scope - KM focuses on exploiting and realizing knowledge from the organization's workforce, fostering a culture where knowledge sharing can thrive, and increase the overall value of intellectual capital required for making decisions. KM is a fundamental part of how the IT department conducts it daily business.

Process Benefits

- Improved efficiency through reducing the need to rediscover knowledge
- Reduced Incident and problem-solving time through sharing of workarounds and previous resolutions
- Reduced design time through sharing of information related to current and past design projects
- Better strategic decisions based on captured and categorized knowledge, rather than institutional memory

- Gain overall efficiency through reuse of previous plans, documents, etc.
- Increase innovation through knowledge sharing and collaboration
- Serve as a process enabler allowing for the organization's knowledge workers to share ideas and collaborate in ways that would not have been possible previously
- Improved project management through knowledge transfer and data availability in existing systems (i.e. allowing data transparency across PM systems)
- Proactively facilitates and rewards knowledge creation, transfer, and utilization

Expected Outcomes

- Support for lifecycle management of data and knowledge assets
- Establishment of content capture and exchange standards
- IT Service knowledge architecture is defined
- Sources of IT Service knowledge and responsible knowledge owners are identified

- Information required to provide Services from external management systems is made available (programs, projects, business plans, Changes, Releases, deployments, Problems, and other knowledge)
- Knowledge is acquired, structured, published and maintained
- Agreed to call resolution rates and the enablement of those resolution rates through provided knowledge solutions is measured, reported and improved

Process Activities

[KM1] Establish Knowledge Management Framework

This activity defines all direction, guidance, policies, and procedures for how the process will be performed. All of this is collectively referred to as the "KM process framework" and is used as reference information for all other activities. Knowledge Management responsibilities are integrated into career paths, job descriptions, and skill requirements. This information is reviewed in the Evaluate Process Performance activity, which generates recommendations for making changes and improvements to the KM process framework.

[KM2] Create and Maintain Knowledge Architecture

The Knowledge Architecture is a framework of policies, standards, and conventions for collection, formatting and organizing process and Service information assets in a consistent manner. This architecture provides a reference model for use in designing and building processes and Services. It also provides a way to define the various segments of KM as the organization matures this process through addressing priorities or weaknesses in KM. The Knowledge Architecture must balance the need to enter and collect information to gain knowledge against a simplistic design that is understandable, usable, and sustainable.

[KM3] Acquire Knowledge Assets

This activity involves all tasks and operations required to harvest targeted information packages that require processing and manufacturing into knowledge assets. These assets are made available through the Service Knowledge Management System (SKMS). Knowledge asset acquisition activities use common processes based on standard data and information models.

[KM4] Analyze Knowledge Assets

Conduct an SME analysis of captured raw knowledge assets, consequential information, and data that has been extracted in the Acquire Knowledge Assets activity. It is envisioned that most knowledge assets will be harvested from authoritative data sources in a consistent format. These knowledge packages will require SME reviews for technical, legal and publication compliance. Once submitted, a rigorous material review process against prescribed submission criteria is performed.

[KM5] Publish and Manage Knowledge Assets

This activity covers all tasks required to make available and deliver knowledge assets to users. It can include both proactively and reactively supplying knowledge.

[KM6] Monitor, Manage and Report Knowledge Management

In this activity, all Knowledge Management activity is monitored to determine whether suitable progress is being made. Unsatisfactory results are reported and may result in actions taken to address any issues.

[KM7] Evaluate Knowledge Management Performance

This activity describes the tasks required to assess the efficiency and effectiveness of the Knowledge Management process. It includes the capture of information, the relationship with other process areas, and the suitability of procedures and training. It is used as a basis to ensure that the Knowledge Management process remains fit for purpose and identifies where changes to the process might be required.

7). Release and Deployment Management (RDM)

Purpose - The purpose of this process is to deploy Releases into the live environment in a controlled manner. Release and Deployment Management ensures that the integrity of the live environment is protected, and correct components are released. This must be in a time frame that meets the business customer's Service needs and does not cause an SLA breach.

Scope - This scope includes the processes, systems, and functions to package, build, test and deploy into the production or live environment for use. RDM establishes the Service as specified in the Service Design Package (SDP) and formally hands the Service over to Service Operations. The package includes all configuration items (CIs) required to implement the release.

Process Benefits

- Minimized disruption of Service to the business partner, due to synchronization of Releases involving hardware and software components from different platforms and environments
- Early Life Support (ELS) becomes part of the process
- Effectively communicates and manages expectations of the business customer during the planning and rollout of new Releases
- Reduction in errors through the controlled release of hardware and software to the live environment
- Unsuccessful deployed Releases are reversed, and environment is recovered
- Enhanced use of resources due to combined efforts when testing new Releases
- Reduction in unplanned work due to better control of Service components and releases
- Overall reduction in configuration variance
- Reduction in unplanned work

Expected Outcomes

- Requirements for Releases are established and agreed upon with affected parties
- Releases of new or changed Services and Service components are planned
- Releases are designed
- Releases are tested prior to deployment
- Approved releases are deployed
- Integrity of hardware, software, and other Service components is assured during deployment of the release
- Unsuccessful deployed Releases are reversed
- Release information is communicated to affected parties

Process Activities

[RDM1] Establish Release and Deployment Management Framework

This activity defines all direction, guidance, policies, and procedures for how the process will be performed. All of this is collectively referred to as the "RDM process framework" and is used as reference information for all other activities. This information is reviewed in the Evaluate Process Performance activity, which generates recommendations for changes and improvements to the RDM process framework.

[RDM2] Plan Release and Deployment Program

This activity determines the approach for how each Release is prepared and the type of deployment that is necessary. The Release planning covers building, testing, and verifying the Release, and establishes a model for how the finalized Release should be deployed.

[RDM3] Design and Build Release

This activity determines what needs to be built for the Release and how it will be assembled and deployed. As a result, the Release build, installation, and rollback scripts are designed at a high level. Software and hardware components are obtained for the build activity and the test environment is created.

[RDM4] Test and Verify Release

This activity tests the built Release Package and determines if installation, configuration, and rollback work properly. Once successful, the Release is ready for deployment. If testing fails, the Release must go through another round of either design or build, and a subsequent re-testing.

[RDM5] Prepare Deployment Capabilities and Perform Transition Administration

This activity administers the transition of assets, resources, knowledge, and anything else that is transferred to or from the IT infrastructure. This ensures that appropriate asset data is provided to the Asset Management process to reflect the transition status. Items impacted include location, financial status (support contracts), and ownership.

[RDM6] Perform Deployment and Activate Service

This activity executes all tasks necessary to complete the actual deployment. In this activity, the capability status moves from "Not Deployed" to "Deployed". This activity verifies the integrity of the solution under deployment and transitions the new changed Service to Operations.

[RDM7] Review and Close Deployment

This activity reviews tasks completed during deployments and determines if all objectives of the deployment plan were met. A management plan is established for outstanding risks, issues, Incidents and Known Errors before the deployment is closed. Deployment is completed with a handover of the support to Service Operations.

[RDM8] Monitor, Manage and Report Release and Deployment Management

In this activity, all Release and Deployment Management activity is monitored to determine whether suitable progress is being made. Unsatisfactory results are reported and may result in actions taken to address issues.

[RDM9] Evaluate Release and Deployment Management Performance

This activity describes the tasks required to assess the efficiency and effectiveness of the Release and Deployment Management process. It includes the capture of information on records, the relationship with other process areas, and the suitability of procedures and training. It is used as a basis to ensure that the Release and Deployment Management process remains fit for purpose and identifies where changes to the process might be required.

8). Service Validation and Testing (SVT)

Purpose - Service Validation and Testing provides evidence that the new/changed Service meets the business customer requirements, including any documented SLAs, thus limiting risk as Changes are introduced into the production environment. The Service is tested explicitly against all parameters in the Service Design Package, including functionality, availability, continuity, security requirements, usability, and regression testing.

Scope - This process focuses on the testing and validation of a fully functional solution that is designed to meet stakeholder requirements and the stakeholder acceptance of that solution prior to roll-out. The scope of Service Validation and Testing is all approved Releases and those components as defined in Release and Deployment Management. It includes all Configuration Items required to implement a Release, and the similar and related systems that make up the production environment.

Process Benefits

- Ensures that Releases meet the criteria for utility and warranty
- Business customer confidence in the success of Releases resulting in elevated satisfaction
- Testing is done from an overall Service perspective, not just for component or system
- Reduction in business customer resources to test releases
- A structured validation and test process that provides evidence that the new or changed Service supports the business requirements as set in Service Strategy
- Ensures that business customer requirements are met as set forth in the Service Design Package

- Improve agility for application Releases and ability to follow release methodologies (Agile, DevOps, etc.)
- Overall reduction in Incidents

Expected Outcomes

- Validation and testing of Services and Service components are planned
- Validation activities and tests are designed
- Only validated, tested and approved Releases are deployed
- Integrity of hardware, software, and other Service components is assured during deployment of the Release
- Releases that fail validation and testing are reversed or remedied
- Validation and testing information is communicated to interested parties

Process Activities

[SVT1] Establish Service Validation and Testing Framework

This activity defines all direction, guidance, policies, and procedures for how the process will be performed. All of this is collectively referred to as the "SVT process framework" and is used as reference information for all other activities. This information is reviewed in the Evaluate Process Performance activity, which generates recommendations for changes and improvements to the SVT process framework.

[SVT2] Oversee Service Solution Testing

This activity is responsible for the testing the Service prior to the introduction of Changes to the environment that affects the Service. SVT is more commonly an iterative process.

[SVT3] Test Solution

Solution testing validates the solution and its features conform to design specifications and requirements prior to deployment. It also verifies that interim work products exist and conform to standards.

[SVT4] Accept Solution

This activity validates that the proposed solution, either as individual artifacts or in its complete form, meets end-user acceptance criteria.

[SVT5] Monitor, Manage and Report Service Validation and Testing

This activity supports continuous monitoring and analysis of operational results data and comparison with Service achievement reporting to identify Service Validation and Testing trends and issues. SV&T information is used to generate detailed Service component reporting as well as a perspective on overall Service Availability.

[SVT6] Evaluate Service Validation and Testing Performance

This activity describes the tasks required to assess the efficiency and effectiveness of the Service Validation and Testing process. It includes the capture of information on records, the relationship with other process areas, and the suitability of procedures and training. It is used as a basis to ensure that the Service Validation and Testing process remains fit for purpose and identifies where changes to the process might be required.

Chapter 9 – Service Operations

The Service Operations phase of the Service Lifecycle controls the full range of matters relating to sustaining assured Service delivery, system and network availability, and information protection for information technology (IT) capabilities that support the provided Services balancing stability with responsiveness. The Service Operations Domain Owner serves as the approval authority to introduce new initiatives, ensures standards-based configuration and operation of all infrastructure, controls the runtime aspects of Services to ensure that Services behave correctly and within SLAs, administers and controls security policies, identifies Incidents and infrastructure issues, performs Problem resolution, and implements metrics that track the overall progress of the Service Operations Domain. The Service Operations Domain Owner must ensure that processes that support Services are executed in a cost-effective manner and measure the effectiveness of controls to determine how well the controls achieved the planned control objectives. As such, processes must have an internal (IT) and external (business customer) focus. It is within this Service Operations Domain that the business customer determines the ongoing value extended by the Service Provider.

There are two key definitions whose difference is more prominent in the Service Operation Domain than the other Service Lifecycle Domains:

Function: A team or group of people with like skill sets and the tools they use to carry out one or more processes or activities.

Process: A structured set of activities designed to accomplish a specific objective.

This difference is notable since, in addition to processes, the Service Operation Domain has several functions (Service Desk, Applications Management, Technical Management, and IT Operations Management). These functions are described in the Functions section below.

> The one key element to remember regarding Service Operation is that this is where the business customers realize the value from the Service Lifecycle. It does not matter how well a Service is strategized and differentiated from competing alternatives, designed, built, tested, or transitioned if the Service does not deliver the results the customer expects. Recommend milestones for process implementation and Service improvements

Implementing and Improving ITSM

1). **Access Management (ACM)**

Purpose - Access Management is the process of granting authorized users the right to use a Service while preventing access to non-authorized users. The process provides the ability to control and track who has access to data and Services ("Who" may be another system, service, or process, as well as an individual or group). It contributes to achieving the appropriate Confidentiality, Integrity, and Availability of the organization's data and includes levels of access to the Service Catalog for requesting Services, access to data, and access to facilities.

Scope - Access Management enables the management of the Confidentiality, Integrity, and Availability of data and intellectual property. This process operates within and enforces controls described by Availability Management, IT security policies and organization directives; control of identities and their associated access rights will vary depending upon the level of access required and the adjudicated risk tolerance of malicious access.

Process Benefits

- Access to Services is aligned with organizational strategy
- Data is protected from accidental and intentional malicious attempts
- Better controlled environment when access needs to be revoked, such as job changes, retirements, and discontinuation of Services
- Employees have the right access to perform their jobs
- Access to data and Services is controlled
- Processes are in place to demonstrate compliance with organization policies
- Consistent enforcement of Service, data, and facilities access

Expected Outcomes

- A definitive source permits the user access to information and Services while unauthorized access attempts receive denial of access
- An accurate identity and rights registry exists that undergoes periodic maintenance and review
- Auditable record of access attempts is maintained and available to authorized personnel

- Data necessary to demonstrate compliance relative to service and information access is available
- Security vulnerabilities and Incidents are identified, monitored and reported
- Unauthorized access to information, applications, and infrastructure is detected, reported, and resolved
- Access-related security Incidents are defined, and access controls are regularly tested

Process Activities

[AcM1] Establish Access Management Framework

This activity defines all direction, guidance, policies, and procedures for how the process will be performed. All of this is collectively referred to as the "AcM process framework" and is used as reference information for all other activities. This information is reviewed in the Evaluate Access Management Performance activity, which generates recommendations for changes and improvements to the AcM process framework.

[AcM2] Evaluate and Verify Access Request

This activity evaluates and verifies the identity of the person listed in each request and verifies that a reasonable substantiation exists for the access to the system, application, or availability to perform specific tasks within an application. This activity also verifies that the request has been approved by the appropriate authority.

[AcM3] Create and Maintain Identity

This activity creates new identity records in the identity database and performs appropriate edits and deletions to existing identity records.

[AcM4] Provide and Maintain Access Rights

This activity provides access rights based on predefined policies, directives, and regulations. It updates the identity records to reflect updated access rights and confirms that access rights have been implemented or revoked. Access rights can be removed as well as granted. Accordingly, this activity will restrict or revoke rights to execute policies and decisions made by the appropriate authority.

[AcM5] Monitor, Manage and Report Access Management

In this activity, Access Management activities are monitored to determine whether suitable progress is being made. Results are reported, and unsatisfactory results may lead to a review of Access Management actions. In addition, responses are provided to requests for information and status of the Access Management process.

[AcM6] Evaluate Access Management Performance

This activity describes the tasks required to assess the efficiency and effectiveness of the Access Management process. It includes the capture of information, the relationship with other process areas, and the suitability of procedures and training. It is used as a basis to ensure that the Access Management process remains fit for purpose and identifies where changes to the process might be required.

2). **Event Management (EM)**

Purpose - The purpose of the Event Management process is to identify and prioritize all Events that occur throughout the IT infrastructure and establish the appropriate response to those Events. Event Management monitors, filters, and notifies of actions and occurrences that influence the Services provided. This process is proactive and reactive. Proactively, Operations is notified of Events that may cause Service degradation and outages enabling operations to take steps necessary to avert any SLA breach. Reactively, Event Management interfaces with Operations, Incident, Problem and Change Management to provide information and corrective actions for those Events.

Scope - Event Management includes occurrences or actions that affect the ability to provide Services. These may be related to: security, performance of CIs, component failure, facilities, capacity, or issues related to compliance and contracts or licensing. Event Management can be used to capture or display near real-time monitoring data enabling increased understanding of the IT Infrastructure and help shape Service Levels. Tool sets are pre-engineered to support automated monitoring and responses to Events, including pre-populating an alarm Event.

Process Benefits

Improved understanding of critical IT Service and infrastructure components and systems

- Decreased labor costs due to automated responses to Events
- Automation for escalating exception conditions to the Incident Management Process to engage automatically, which improves Service Availability.
- Higher productive staff through reduction of monitoring non-consequential Events
- Ability to preclude Incidents, increasing availability of Services
- Quicker Return to Service due to notification from the source of the Event
- Standardization in Event notification, enables better responses from Operations
- Reduction in the TCO for monitoring resources
- Monitoring of IT should be Service- and business-focused

Expected Outcomes

- Improved understanding of critical IT Service and infrastructure components and systems
- Automation for escalating exception conditions to the Incident Management Process to engage automatically, which improves Service Availability.
- Enhances ability to make informed decisions based on business needs
- Defined warning criteria, used to display alerts in network monitoring tools, enabling improved visibility into potential Service disruptions and allowing decisions/actions to proactively lessen potential impacts

Process Activities

[EM1] Establish Event Management Framework

This activity defines all direction, guidance, policies, and procedures for how the process will be performed. All of this is collectively referred to as the "EM process framework" and is used as reference information for all other activities. This information is reviewed in the Evaluate Process Performance activity, which generates recommendations for changes and improvements to the EM process framework.

[EM2] Define and Log Requirements

This activity involves receipt of all predefined Events detected into the Event Management System (EMS) monitoring the IT environment. When an Event is actively or passively detected, it is the responsibility of those managing the device to ensure that the Event is defined and logged in an agreed format and that protocol is adhered to for handling the EMS.

[EM3] Filter Event

This activity determines if the Event must be communicated or ignored based on predefined criteria.

[EM4] Correlate Event

This activity describes the tasks involved in reviewing Service Requests that were fulfilled in this activity. The organization's predefined business goals are applied to significant Events to determine what actions are required. Events are correlated by the EMS to determine commonalities and appropriate response action.

[EM5] Trigger Response

After an Event is detected, filtered and correlated, the appropriate and specific Event notification and response activities are initiated. The response includes opening an Incident, changing the status or severity of an Event, dropping an Event, or sending the Event for automated recovery.

[EM6] Execute Auto Response

In this activity, a pre-defined automated response is initiated by the EMS (e.g. rebooting and/or restarting a device, initiating a batch job, etc.). These responses do not require human intervention.

[EM7] Generate Alert

This activity identifies those Events requiring human intervention and provides necessary information to determine appropriate action. Additionally, this activity transmits the Event information to Incident Management, which manages routing/escalation to the proper level for resolution.

[EM8] Clear Event

In this activity, the status of the Event is confirmed as cleared and appropriate updating of Event records is made.

[EM9] Monitor, Manage and Report Event Management

In this activity, all Event Management activity is monitored to determine whether suitable progress is being made. Unsatisfactory results are reported and may result in actions taken to address any issues.

[EM10] Evaluate Event Management Performance

This activity describes the tasks required to assess the efficiency and effectiveness of the Event Management process. It includes the capture of information on records, the relationship with other process areas, and the suitability of procedures and training. It is used as a basis to ensure that the Event Management process remains fit for purpose and identifies where changes to the process might be required.

3). Incident Management (IM)

Purpose - The purpose of Incident Management is to restore normal Service operation as quickly as possible and minimize the adverse impact on business customer operations, thus ensuring that the best possible levels of Service quality, security, and availability are maintained. The focus is on reducing the duration and consequences of Service outages from a business customer perspective; not on finding the root cause of the Incident. (Root cause will be in scope for Problem Management below).

Implementing and Improving ITSM

Scope - The scope includes any disruption or potential disruption of Service. The process allows for three different paths: Normal, Major, and Security-related. Defining a Major Incident is an important aspect of Incident Management process definition. Those Incidents that have the highest impacts and are most disruptive to the affected Service components must be managed in a separate subprocess. The impact and urgency thresholds are agreed upon in a Priority Matrix, which includes all levels of Incident priority (see IM3 below). Additionally, security Incidents require a separate sub-process since these tend to occur because of activities intended to disrupt or degrade Services, rather than because of human error or material failures. These have different reporting criteria and may or may not adversely affect one or more Services. Other Incidents are considered normal.

Process Benefits

- Ability to detect and resolve Incidents more efficiently, which results in greater availability of the Service to the business customer.
- Ability to align IT activity with real-time business priorities. Incident Management includes the capability to identify business customer priorities and dynamically allocate resources as necessary.

- Identification of potential improvements to Services.
- Potential to identify needed Service or training during the handling of Incidents
- Improved information flow to business customers regarding Service restoration
- Basis of information for Problem Management.
- A single Incident Management process for use across the organization
- Better focus on restoring Service as opposed to just performing Root Cause Analysis (RCA)
- A searchable base of Incidents and workarounds to better resolve Service outages
- A standard method of prioritization, categorization, and escalation of Incidents
- Transparency into the status of Incident resolution
- Incident models to allow for more efficient resolution of Incidents

Expected Outcomes

- Incidents are recorded and categorized
- Incidents are prioritized and analyzed
- Incidents that have not progressed according to accepted Service Level timelines and thresholds are escalated
- Incidents are resolved and closed

- Major incidents are reviewed
- Information regarding the status and progress of reported Incidents is communicated to interested parties

Process Activities

[IM1] Establish Incident Management Framework

This activity defines all direction, guidance, policies, and procedures for how the process will be performed. All of this is collectively referred to as the "IM process framework" and is used as reference information for all other activities. This information is reviewed in the Evaluate Process Performance activity, which generates recommendations for changes and improvements to the IM process framework.

[IM2] Identify, Report and Log Incident

The Incident is identified and logged by the Service Desk resulting in the creation of an Incident record.

[IM3] Categorize and Prioritize Incident

The Incident is categorized and prioritized. Categorization is based on the systems, applications, Service affected, or the requestor's business support role. Prioritization is based on urgency and impact. The record is assigned to an analyst for diagnosis and investigation.

The path and procedures involved are based on how the Incident is categorized, for example, Normal, Major, or Security, etc. If the Incident is categorized as a request for Service, it is transferred to the Request Fulfillment process as a Service Request. Incidents exceeding a defined threshold of impact and urgency are categorized as Major Incidents and appropriate procedures are invoked.

[IM4] Investigate and Diagnose Incident

Incidents and all associated data are accessed to identify appropriate responses and actions and to formulate Incident Resolution Plans. Actions may include identifying workarounds, re-categorizing the Incident based on further analysis, and updating Incident records.

[IM5] Resolve and Recover Incident

Actions necessary to resolve the Incident and restore Service are executed. Resolution and restorations may be in the form of existing workaround solutions, or alternatively creating a Request for Change to implement a new solution. It also prompts any action necessary to recover the Service to approved Service Level Agreements (SLA), Operational Level Agreements (OLA) and/or Underpinning Contracts (UC).

[IM6] Close Incident

This activity ensures that all required Incident documentation is complete, including details of cause, expended effort for resolution, and outcome. A review of the Incident's original categorization against available root-cause information is used to determine categorization accuracy. This activity obtains stakeholder agreement with resolution activity and status.

[IM7] Monitor, Manage and Report Incident Management

This activity supports continuous monitoring and analysis of operational results data and comparison with Service achievement reporting to identify Incident Management trends and issues. Incident Management information is used to generate detailed Service component reporting as well as a perspective on overall Service Availability.

[IM8] Evaluate Incident Management Performance

This activity describes the tasks required to assess the efficiency and effectiveness of the Incident Management process. It includes the capture of information on records, the relationship with other process areas, and the suitability of procedures and training. It is used as a basis to ensure that the Incident Management process remains fit for purpose and identifies where changes to the process are required.

4). Problem Management (PM)

Purpose - The purpose of Problem Management is to prevent Problems and Incidents from happening, to eliminate recurring Incidents, and to minimize the impact of Incidents that cannot be prevented. Problem Management includes the activities required to diagnose the root cause of Incidents, determining the resolution to those Problems and providing workarounds to Incident Management.

Scope - A Problem is defined as a cause of one or more Incidents. The cause is not usually known at the time a Problem record is created, and the Problem Management process is responsible for further investigation. Problem Management has aspects of both reacting to Problems and proactively identifying and solving Problems and Known Errors before more Incidents occurs.

Problem Management finds trends in Incidents, groups those Incidents into Problems, identifies the root causes of problems, and initiates Request for Change (RFC) against those Problems. Also, PM maintains information about Problems and the workarounds and resolutions to reduce the number and impact of Incidents over time and proactive PM may also inform Continual Service Improvement. This process has a strong interface with Knowledge Management and tools such as the Known Error Database (KEDB).

Although Incident Management and Problem Management are separate processes, they are closely related and typically use the same tools, and have similar categorizations, impact, and priority coding systems. This ensures effective communication when dealing with Incidents and Problems that are related. Problem Management also ensures that the resolution is implemented through appropriate control procedures such as Change Management and Release and Deployment Management.

Process Benefits

- Incident trends are identified and proactively investigated as a Problem
- Higher availability of IT services
- Higher productivity of business and IT staff
- Reduced expenditure on workarounds or fixes that do not work
- Reduction in cost of effort in fire-fighting or resolving repeat Incidents
- Reduces the chance of having to invoke the Business Continuity Plan
- A Known Error Database (KEDB) reduces time to resolution and allows learning from historical data
- A structured process based on prioritization schemes to allocate resources for solving Problems

- Ability to distinguish between restoring Service (IM) and Root Cause Analysis (PM), which will create greater availability of Services
- Better integration of the supporting processes.

Expected Outcomes

- Problems are identified, recorded and classified
- Problems are prioritized and analyzed
- Problems are resolved and closed
- Problems not progressed according to defined Service Levels are escalated
- The effect of unresolved Problems is minimized
- The status and progress of the resolution of Problems are communicated to stakeholders

Process Activities

[PM1] Establish Problem Management Framework

This activity defines all direction, guidance, policies, and procedures for how the process will be performed. All of this is collectively referred to as the "PM process framework" and is used as reference information for all other activities. This information is reviewed in the Evaluate Process Performance activity, which generates recommendations for changes and improvements to the PM process framework.

[PM2] Identify and Log Problem

This activity ensures that problems are identified through resource monitoring, trend recording, and analysis.

[PM3] Categorize and Prioritize Problem

Problems are classified to support active analysis, Problem resolution, and post-Problem forensic review. This activity also classifies Problem severity and potential impact to enterprise operations and goals.

[PM4] Investigate and Diagnose Problem

This activity includes Root Cause Analysis, creating workarounds, and recording Known Errors. If a workaround is identified and approved for deployment, this activity ensures that the workaround is known to be effective, and sufficient evidence exists to support the Root Cause Analysis. A Known Error Record is created or updated that describes Problem diagnosis and lists available approved workarounds. This activity also updates the Problem record to indicate the diagnosed Problem.

[PM5] Resolve Problem

This activity includes the search for a solution, steps planned to implement the solution and eliminate Known Errors, and tracks infrastructure Changes. Once the resolution has been documented in the Problem and Known Error records, the activity culminates in Request for Change or Project Proposal submissions.

[PM6] Close and Review Problem

The Problem record is closed, Known Error records are updated, and major Problems are reviewed for performance quality, process adherence, and lessons learned. Prior to closing, each Problem record is checked to ensure completeness and accuracy of detail. Major Problems are reviewed, and results are disseminated through enterprise communication (including extended enterprise stakeholders such as vendors), staff training, and Service review.

[PM7] Monitor, Manage and Report Problem Management

This activity ensures that all Service Requests are effectively and efficiently managed throughout the process lifecycle. All request data and status changes are examined for consistency and recorded in Problem Management records.

[PM8] Evaluate Problem Management Performance

This activity describes the tasks required to assess the efficiency and effectiveness of the Problem Management process. It includes the capture of information, the relationship with other process areas, and the suitability of procedures and training. It is used as a basis to ensure that the Problem Management process remains fit for purpose and identifies where changes to the process are required.

5) Request Fulfillment (RF)

Purpose - The purpose of the Request Fulfillment process is to fulfill Service Requests from users and route each Request to the appropriate process for handling within accepted Service Levels. Request Fulfillment is responsible for the entire lifecycle of the Request.

Scope - Request Fulfillment encompasses fulfillment of Service Requests within agreed Service Levels. Requests can come from a business customer by direct communication or automated menu system. This process interacts at the process framework level of other specific processes to determine which types of Service Requests should be handled by which processes, e.g., Request for Changes interacts with the Change Management process. Request Fulfillment is responsible for the entire lifecycle of the request.

Process Benefits
- Service improvement through repeatable and measured fulfillment
- The Service Desk better prioritizes requests by separating Incidents from Service Requests
- Business customers have quick and easy access to standard Services
- Standard process for financial approval of standard Service Requests

Expected Outcomes

- Service Requests are recorded and classified
- Service Requests are prioritized and analyzed
- Service Requests are fulfilled and closed
- Service Requests that have not progressed according to accepted Service Level timelines and thresholds are escalated
- Information regarding the status and progress of Service Requests is communicated to interested parties

Process Activities

[RF1] Establish Request Fulfillment Framework

This activity defines all direction, guidance, policies, and procedures for how the process will be performed. All of this is collectively referred to as the "RF process framework" and is used as reference information for all other activities. This information is reviewed in the Evaluate Process Performance activity, which generates recommendations for changes and improvements to the RF process framework.

[RF2] Log and Validate Service Request

All reported Service Requests must be logged in a ticket management system with relevant details (user contact information, asset information, etc.) and are categorized, prioritized and assigned to the appropriate team for fulfillment. If the Request does not meet established criteria for fulfillment, it is rejected, and the user notified.

[RF3] Fulfill or Route Service Request

The Service Request is analyzed to determine the appropriate team to perform the fulfillment activities. If the Request is resolved within Request Fulfillment, the user is contacted to verify resolution. Upon user satisfaction, The Service Request record is updated and closed. If the Request is transferred to another process, all relevant information and documentation are routed to the appropriate team and the receiving process is notified about the assigned Request item. Request Fulfillment retains ownership of the Service Request and tracks fulfillment progress through user acceptance and closure.

[RF4] Obtain Acceptance and Close Service Request

This activity examines the work history of a Service Request with a "Resolved" status. It ensures that all required documentation is complete, including resolution details, effort expended and outcome. A review of appropriate classification and prioritization is conducted and stakeholder agreement with resolution activity and status is obtained for formal closure.

[RF5] Monitor, Manage and Report Request Fulfillment

This activity ensures that all Service Requests are effectively and efficiently managed throughout the process lifecycle. All request data and status changes are examined for consistency and recorded in Service Request records.

[RF6] Evaluate Request Fulfillment Performance

This activity describes the tasks required to assess the efficiency and effectiveness of the Request Fulfillment process. It includes the capture of information, the relationship with other process areas, and the suitability of procedures and training. It is used as a basis to ensure that the Request Fulfillment process remains fit for purpose and identifies where changes to the process might be required.

Service Operation Supporting Functions

A function is described as a team, organization unit or group of people that perform certain activities or types of work. They typically have the similar skill sets and resources to carry out their duties to achieve specific outcomes. The function is responsible for defining the standards and procedures to be followed when operating within the function. It is a challenge to include functions in this framework, as these functions already exist, have been in existence for some time, and already have processes and procedures with some level of effectiveness. The benefit gained by documenting the functions is the understanding of standardized integrated processes and how functions fit into the overall IT Service Management Framework as services are delivered to IT business customers.

Service Operation Function Roles and Responsibilities

While all functions have unique roles and responsibilities, they also have two roles in common: Function Owner and Function Manager.

The **Function Owner** has the following responsibilities:

- Ensure that the function's design includes the policies of the processes performed as part of the function
- Ensure that the function follows the governance guidelines of the appropriate governing bodies
- Ensure the integration of various processes utilized in the function
- Coordinate with the various Service Lifecycle Domain Owners with relation to their processes
- Coordinate with Process Owners to suggest improvements to their processes

The Function Owner must have the authority to direct members across Service Lifecycle Domains. Therefore, Function Owners must be in senior leadership.

The **Function Manager** is responsible to the Function Owner and performs day-to-day operational and managerial tasks demanded by the function. The Function Manager does not necessarily fall into the Function Owner's organizational chain of command. The Function Manager has the following responsibilities:

- Monitor the function, using qualitative and quantitative Key Performance Indicators (KPI) and make recommendations for improvement
- Play a key role in developing requirements and maintaining the function's tools
- Escalate questions related to the function
- Identify training requirements for all support staff and ensure that proper training is provided to meet the requirements
- Provide metrics and reports to leadership and business customers in accordance with outlined procedures and agreements

1). **Service Desk (SD)**

Purpose – As the single, primary point-of-contact, the Service Desk is the interface between the user and the Service. If there is an issue whether it is an unclear Event or Alert message, an Incident or Problem, or an Access issue, the user is going to contact the Service Desk for assistance if the issue cannot be resolved through self-help methods.

The purpose of the Service Desk is to:

- Be primary contact point for all calls, questions, Service Requests, complaints, and remarks
- Be primary provider of ongoing monitoring and management of business customer satisfaction through appropriate communication channels
- Manage the Incident lifecycle
- As other processes mature, the Service Desk becomes more involved in areas such as Configuration Management, but the primary purposes above remain the same.

Scope - There are different concepts of operation environments established and therefore differing scopes for Service Desk support. The scope of the Service Desk is also sometimes determined by the Service Level Agreements (SLAs) that were defined during business customer negotiations.

Benefits and Expected Outcomes

- Business customer satisfaction – the business customer is generally better served and better satisfied by establishing a single point-of-contact for all Incidents and Service Requests
- Decreases in overall business impact of Incidents – Incidents are handled more efficiently through the Service Desk due to consistent use of process, procedures, and tools for resolution
- Cost reduction – Service Desk reduces duplicative efforts through better communication and shared knowledge during Incident resolution
- Better communication – Single point of information flow ensures consistency in reporting to decision-makers during Service outages
- Reduction in redundancy and better implementation of global solutions through greater knowledge sharing
- Increased business customer satisfaction through quicker resolution, better communication, and stricter adherence to SLAs
- Reduction in Service outages and overall time to restore Services

Implementing and Improving ITSM

Relationship to other Functions

- **Application Management** - Provides second-tier support during Incident resolution, especially as it relates to the business applications and systems
- **IT Operations Management** - Provides second-tier support during Incident resolution, especially as it relates to the Services monitored through operations. Will relay to Service Desk any issues related to productions activities, job scheduling, and operational failures
- **Technical Management** – Provides second-level support during incident resolution, especially as it relates to the IT infrastructure

Relationship to Processes

- **Access Management** - Provides support in granting access to business partners and users
- **Change Management** – Participate in Change Advisory Boards (CABs) and track Changes against Incidents
- **Configuration Management** – Service Desk personnel may also be designated to fill the role of Configuration Management Librarian if appropriately trained

- **Event Management** – Provides support when there is an Event
- **Incident Management** – Primary executer of the Incident Management process and procedures. Takes ownership of all Incidents
- **Problem Management** – Participates in Problem resolution and trend analysis
- **Request Fulfillment** – Primary point-of-contact for coordinating business customer Requests for Services as well as informational Requests
- **Service Level Management** – Performs requests in conjunction with established SLAs and OLAs. Service Desk monitoring is also used to inform SLM of SLA breaches

2). Application Management (APP)

Purpose - The purpose of Application Management is to support the lifecycle of applications (requirements, development, build, deployment, operation, optimization, and retirement) that enhance organization's IT ability to provide Services to its business customers.

Scope - The scope of Application Management encompasses the lifecycle of all applications, provided by a vendor or developed in-house, from understanding the strategic goals of IT and its business customers to the retirement or discontinued use of the application.

This includes ensuring that the applications sustain the Services in the production environment. Basically, this covers any software, other than operating systems and firmware that resides on the IT Infrastructure in support of Services provided to the business customers.

Benefits and Expected Outcomes
- Much greater focus on software and software development in support of Service
- More efficient Incident resolution related to applications issues
- Reduced cost of Service implementations through better control of application Configuration Items
- Better visibility into the Application Management processes resulting in lower risks and higher quality
- A standard set of processes for deployment of software
- Centralized control of software licensing, control of software versioning, and the management of a definitive media library (DML)
- Better applications portfolio management through centralization and consistent analysis of software options and consolidation of software options
- Better at meeting cost, quality, schedule, and performance goals

Relationship to Other Functions

- **IT Operations** - Provides Event Management requirements to IT Operations, and guidance on operational management of the technology
- **Service Desk** – Coordinates with Application Management on any issues related to application Incidents
- **Technical Management** – Provides specifications to Technical Management for applications performance configuration and storage requirements

Relationship to Processes

- **Capacity Management** – Determines resources required from infrastructure components for Capacity Management
- **Configuration Management** – Records application CIs and changes to CI fields as they relate to deployment
- **Event Management** – Provides required alerts for IT Operations
- **Incident Management** – Provides resources for resolution of application Incidents

- **Information Security Management** – Identifies security information for access to vendor software, and for any security bypasses used by purchased software, especially software used for monitoring
- **IT Service Continuity Management** – Plans for recovery services as related to applications
- **Problem Management** – Provides resources for resolution of application problems
- **Service Level Management** – Provides performance throughput information
- **Service Portfolio Management** – Provides requirements and prioritization of application development efforts
- **Service Validation and Testing** – Provides testing of Changes and Services as related to applications, as part of the required segregation of duties.
- **Strategy Generation Management** - Provides requirements, needs, and wants from stakeholders
- **Supplier Management** – Supports and often manages suppliers of software

3). IT Operations (ITOP)

Purpose - The purpose of IT Operations is to ensure alignment of three primary areas of responsibility within the IT and business Services. These areas are:

- Management of the day-to-day activities supporting the IT infrastructure
- Operations Control
- Facilities Management

Scope - IT Operations encompasses all computer hardware (owned or deployed on behalf of the organization), all software that runs on the infrastructure, and all network components. It includes monitoring and reacting to Events that impact the environment, job scheduling, performing backups and restores, output management and IT Service Continuity Management (ITSCM) activities. Cloud environments, Command Centers, Data Centers, and outsourced facilities fall within the management scope of IT Operations.

Benefits and Expected Outcomes

- Continuous monitoring of Services - Outages and performance degradation of Services can be mitigated more efficiently through earlier detection
- Better conflict resolution – Centralized operations eliminates conflicting computer resources through holistic scheduling of Events in the environment
- Cost reduction – IT Operations construct reduces duplicative efforts
- Better communication – Single point of information flow ensures consistency in distribution of changing operational priorities
- Business partners will better align IT Operations with organizational goals
- IT Operations should have SLAs, recognized cost of Services, and agreement on requirements for moving new and changed Services into production
- IT Operations is represented throughout the Service Lifecycle

Relationship to Other Functions

- **Application Management** - Provides guidance to **IT operations** about how best to carry out the ongoing operational management of applications
- **Service Desk** – Coordinates with IT Operations on any issues related to productions activities, job scheduling, and operational failures
- **Technical Management** – Provides technical knowledge and expertise related to managing the IT infrastructure

Relationship to Processes

- **Access Management** – Provides support in granting access to business customers and users as it relates to facilities
- **Change Management** – Coordinate maintenance and modifications to infrastructure components
- **Configuration Management** – Responsible for recording new infrastructure CIs and changes to CI fields, such as location
- **Event Management** – Primary human responder to alerts

- **Incident Management** – Monitors for, records, and coordinates with the Service Desk for all infrastructure and job scheduling Incidents
- **Information Security Management** – Assigns facilities and physical security processes and policies. Designs Event Management parameters related to security breaches
- **IT Service Continuity Management** – Primary executer of ITSCM plans. Primary tester of ITSCM plans
- **Problem Management** – Participates in Problem resolution and trend analysis
- **Service Level Management** –Monitors and reports on Events as related to SLAs. Owns recovery of infrastructure within SLA parameters. Function is prominent in OLAs

4). **Technical Management (TechMGT)**

Purpose - The purpose of Technical Management is to provide expertise and knowledge to build and maintain the infrastructure throughout the lifecycle of Services to include design, testing, implementation, operations, and retirement. Technical Management ensures that the organization has access to the right resources to manage technology in alignment with business objectives and strategic goals.

Scope - The scope of Technical Management covers the lifecycle of the infrastructure, from understanding the strategic goals of IT and its business customers, to the design, construction, and testing of Changes in support of the business, to the deployments into production, while ensuring that the infrastructure supports the Services in the production environment. Technical Management will take a role in the event a Service is discontinued or the infrastructure to maintain Service Levels requires upgrades or replacement to infrastructure components.

Benefits and Expected Outcomes

- Individual infrastructure Service components are managed more effectively because staff are adequately trained and skilled
- More efficient Incident resolution related to infrastructure issues
- Reduced cost of Service implementations through better control of infrastructure Configuration Items
- Better Financial Management and understanding of the relationship of high-cost infrastructure to Service.
- Structure should exist to set goals and plans for the technical department in business expertise and technology

- Establishment of training programs to move technicians into management

Relationship to Other Functions

- **Application Management** – Provides specifications to technical management for infrastructure support requirements.
- **IT Operations** - Provides Event Management requirements to IT Operations, and guidance on operational management of the technology
- **Service Desk** – Coordinates with Technical Management on any issues related to infrastructure Incidents

Relationship to Processes

- **Business Relationship Management** - Provides requirements, needs, and wants from stakeholders
- **Capacity Management** – Determines resources required from infrastructure components for Capacity Management
- **Configuration Management** – Responsible for recording new infrastructure CIs and changes to CI fields as they relate to infrastructure upgrades
- **Event Management** – Provides required alerts for IT Operations
- **Incident Management** – Provides resources for resolution of infrastructure Incidents

- **Information Security Management** – Identifies security information for access to infrastructure components and firmware
- **IT Service Continuity Management** – Plans for recovery of new and changed Services as related to infrastructure
- **Problem Management** – Provides resources for resolution of infrastructure Problems
- **Service Level Management** – Provides performance throughput information.
- **Service Validation & Testing** – Provides testing of Changes and Services as related to infrastructure, as part of the required segregation of duties
- **Supplier Management** – supports and often manages suppliers of infrastructure components

Chapter 10 – Continual Service Improvement

It is imperative to be constantly looking for ways to improve Services. With technology changing so quickly, and providing more and improved features, it is necessary to improve Services not only to gain a competitive edge, but in many instances simply to better serve the business partner and provide better solutions with higher availability.

CSI combines techniques, practices, principles, and methods from quality management and Change Management to achieve improvement in processes, Service delivery and quality.

CSI Metrics are actionable measures for decisions related to improving the performance of a Service, process and guiding resource allocation. Metrics must be viewed in an overall context of this Framework. Metrics are unusable if in a silo. As Services and processes are improved, current metrics are reviewed and analyzed to ensure that continued or newly developed measures are in place.

Purpose - CSI combines practices, methods, and principles from quality management and Service measurement practices. The purpose of CSI is to ensure that Services provided within IT remain aligned with business objectives and the ever-changing needs of the consumer. This must be accomplished in concert with improving and maintaining quality and performance of Services.

Scope - The scope of CSI encompasses all Services, internal and external, and the entire lifecycle of IT Service Management processes that support those Services.

Benefits and Expected Outcomes

- Service Owners and Process Owners have a process for understanding ways to improve their areas of accountability
- Improved Return on Investment (ROI) and Total Cost of Ownership (TCO)
- Quality of Services improves
- Quicker recognition of performance issues, allowing for less costly resolutions
- Ensures that Services remain aligned with business objectives
- Better information for planning

- A standardized method for measuring Services and processes
- Standardized process for monitoring and reporting on technology metrics, process metrics, and Service metrics

Continual Service Improvement (CSI) seeks to *continually align and realign IT Services to the changing business needs by identifying and implementing improvements to IT Services that support business processes.*

According to ITIL©, CSI has five objectives:
1. *Reviews, analyze, and make recommendations on improvement opportunities in the entire Service Lifecycle (Service Strategy, Service Design, Service Transition, and Service Operation).*
2. *Review and analyze Service Level Achievement results*
3. *Identify and implement individual activities to improve IT Service quality and improve the efficiency and effectiveness of enabling ITSM processes*
4. *Improve cost-effectiveness of delivering IT Services without sacrificing customer satisfaction*
5. *Ensure that applicable quality management methods are used to support continual improvement activities*

Implementing and Improving ITSM

The Continual Service Improvement (CSI) Model is effective in helping organizations and their leaders have a plan for continuous improvement. The illustration below shows a constant cycle of improvement. As you read through the model, consider how metrics and measures are the focal points in all but the first step.

Metrics and measures are the keys to improvement. How do we know if something needs improvement or if we made an improvement if there are no supporting measures and metrics?

It is also through these measures and metrics that an organization justifies the improvement opportunity. Every Continual Service Improvement opportunity must be justifiable.

CSI interfaces with all four of the other Service Lifecycle phases. The Service improvements help the Service Provider provide a better Service at a better cost, with better security and continuity, delivered in a more risk-averse manner.

CSI interfaces with Service Strategy in a variety of ways. For example, there are opportunities for differentiation of Services, regulatory or security drivers, or even the introduction of a balanced scorecard.

In Service Design, CSI can aid in ensuring alignment between the Service Provider and the business customer. The Warranty processes (Availability Management, Capacity Management, IT Information Security, and IT Service Continuity Management) are the main Service Design CSI targets. Improvements in these four areas enable the Service Provider to align and re-align with the business customer.

In Service Transition, CSI uses measures and metrics in the form of KPIs and CSFs to identify automation targets and opportunities to lessen the risk to the environment upon implementing a new or changed Service.

In Service Operation, CSI is vital to improving the uptime or cost performance of the Service. The Service Operations processes and functions are concerned about uptime, quick restoration after an outage, and ensuring that the business customer can order Services and receive the appropriate Access.

Since the output of one process – or Service Lifecycle phase – is the input to the next, these handoffs offer great opportunities for improvement. Many times, the introduction of acceptance and exit criteria will help close gaps in a methodical way.

Implementing and Improving ITSM

What is the vision? What is the vision in context of the high-level business objectives. Must align with the business and IT strategies

Where are we now? Assess current situation to obtain an accurate, unbiased view of current status of organization. This becomes the baseline assessment for further analysis of business, organization,

Where do we want to be? Understand and agree on priorities for improvement based on deep udnerstanding of the vision. The full vision may be years away but this step provides specific goals and a

How Do we get there? A detailed CSI plan is developed to achieve highe quality service provision by implementing ITSM processes.

Did we get there? Using metrics and measurments, verify if milestones were achieved, high level of process compliance, and business objectives and priorities were met by the level of service

CSI can use the Deming Cycle to control and manage quality. The Deming cycle looks to produce slow and steady improvement.

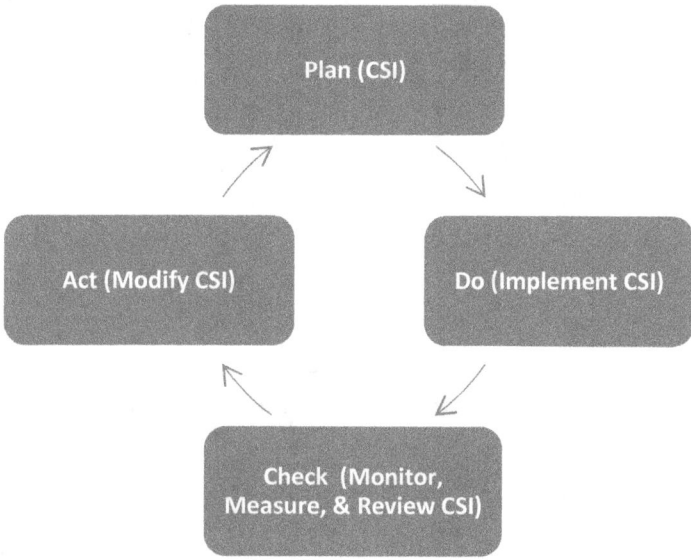

Service Improvements are governed by the improvement lifecycle. The improvement lifecycle is modeled on the Deming Cycle (above). The model establishes a clear pattern for continual improvement efforts.

In the CSI context, baselines establish a starting point for improvement to (a) determine if a Service or process needs to be improved and (b) something we can measure progress against. Sometimes, a baseline is called a benchmark, the recorded state of something at a specific point in time.

There are four main reasons to measure:
1. Intervene – Many times, the metrics and measures show that action is required in a timely manner.
2. Direct – This is the driver for most metrics and measures. Organizations need to set direction and the numbers support the decision.
3. Justify – Metrics offer factual proof that action is either required or not.
4. Validate – Validation of previous decisions requires metrics and measures.

ITIL© gives an idea that bears consideration. Baselines must be agreed-upon, so the starting point is a common understanding. It also needs to be documented and recognized. Lastly, if there is no baseline, then your first measurement becomes the baseline. The baseline or benchmark offers a basis for future metrics and measures to be based.

Section Three

Chapter 11 – Quality, Performance & Metrics

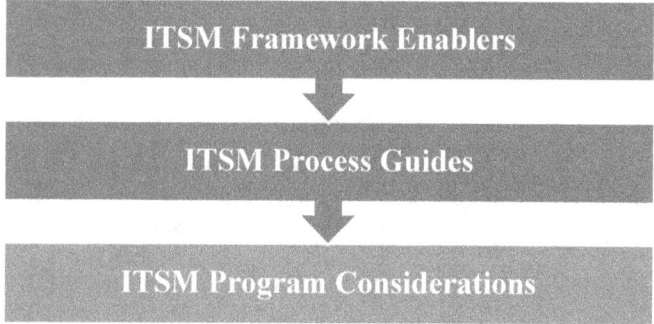

To implement a viable ITSM Program, we need to understand how to measure quality and performance. Service Quality Management (SQM) is a mechanism to define, design, implement, and assess the quality and performance elements of your ITSM Program. SQM is based on a four-phase approach that enables consistent visibility into the business perspective of IT performance, with the mechanisms in place to include the ITSM Program and support future IT investment decisions. This approach is sometimes called the "Deming Cycle" to specifically address IT service quality. We introduced this approach above but are elaborating on a Service Quality focus.

11.1 Service Quality Management

For each of the phases, the activities and outcomes are represented in detail below.

The four phases are the same as the Deming Cycle discussed above:

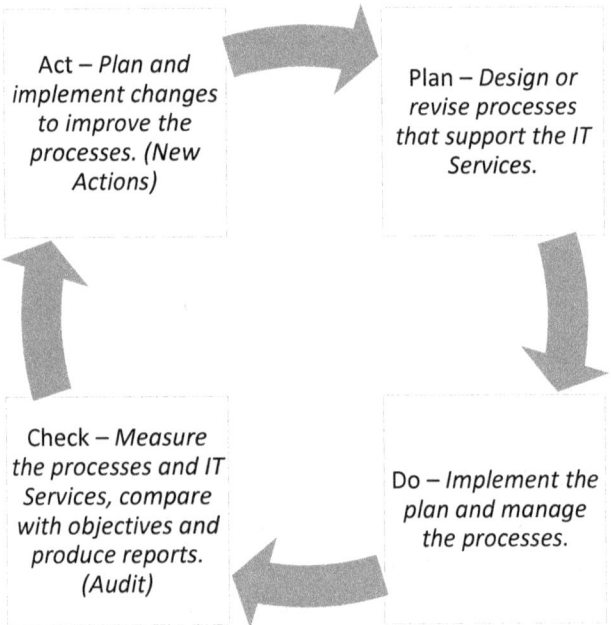

> **The Deming Cycle adapted for Service Quality Management**
>
> - Plan: The quality approach to designing processes that support the IT services
> - Do: Implement the plan to execute the quality approach
> - Check: Measure and monitor the quality management
> - Act: Continuously improve quality management

Plan

Activities: Plan quality scope and objectives; Develop a Service Quality approach and framework; and Communicate scope and objectives

Outcomes: Service Quality purpose, scope, roles, and objectives

Do

Activities: Establish Quality Performance Metrics and Measures; Plan Process Capability Assessments

Outcomes: IT Performance Management Guidance; Process Capability Assessment

Check

Activities: Assess Process Capability; Collect and Report Quality Performance Metrics, Review Quality Performance Metrics

Outcomes: Performance Management and Metrics assessment and evaluation; Process Capability Assessment

Act

Activities: Improve Service Quality; Improve management capability, Plan process capability improvement, track improvement performance

Outcomes: Performance Management improvement; Process Capability improvement

Step One - Plan

The purpose of the Plan step is to establish the objectives and processes necessary to deliver the desired results in accordance with customer requirements and organizational policy and objectives. The planning phase will consist of the developing one deliverable: The Service Quality Plan:

> **Service Quality Plan.** The plan will establish the approach and framework for managing Service quality, and will identify the quality scope, objectives and roles. Service and Process Owners and Managers will plan Service quality specific activities to monitor and address service performance. Service Level Management, Service Owners and Service Managers are responsible to perform Service quality metrics activities related to specific Services. The Service Managers and Process Managers will establish a Performance Management Plan to measure effectiveness of Service

Step Two – Do

Establish and execute the quality approach

The purpose of the step is to implement the quality management processes and capabilities related to (a) Service quality performance metrics AND (b) ITSM process capability improvement

a). Service Quality Performance Metrics

If an organization desires to monitor Service quality and performance, metrics need to be established by:

1. defining the Service metrics
2. defining the measures and data to be collected and assessed
3. defining Service metrics reporting accountability and responsibility

```
                    ┌─────────────────┐
                    │ Ensure Services │
                    │ are defined with│
                    │ enough details  │
                    │ to enable Service│
                    │   measurement   │
                    └─────────────────┘
```

- Ensure Services are defined with enough details to enable Service measurement
- Service Owners and Managers measure Service performance based on Service descriptions and requirements. Service Owners map Service performance metrics to user descriptions and
- Process Owners assess Service Management capabilities to manage Services
- Evaluate and improve the Service quality performance metrics

b). Service Management Capability Improvement

> A process capability improvement model and approach should be defined to support Continuous Service improvements. The approach will guide Process Owners and Managers in their individual continual improvement. This approach takes a practical, phased approach to capability improvement rather than trying to adopt all Service Management best practices in a single step.

Step Three – Check

Monitor and report on Service Quality

This phase executes the monitoring and reporting activities established in the Plan phase. The purpose of this phase is to monitor and measure the processes and Services against policies, objectives, and requirements and report the results to stakeholders.

Below is a five-phase process for monitoring and reporting on Service Quality:

a). Collect and Record Data - This activity will focus on collecting Service quality and process capability data, evaluate data quality and recommend improvement opportunities. The order of activities is:
1. Review outcomes, recommend appropriate updates to process and Service documentation (e.g., Service performance metrics, etc.)
2. Collect information regarding decision-making capabilities, limitations, and measurement
3. Review desired data, data collection practices, and data sources

b). Evaluate Collected Data - Evaluation activities include the following:

1. Evaluate Service and process quality against requirements and develop recommendations
2. Evaluate process capability assessment results and develop capability improvement recommendations
3. Evaluate ITSM quality capabilities and develop recommendations
4. Report Service quality and management capability

c). Conduct Process Capability Assessments - A typical order of events in conducting a process capability assessment is as follows:
1. Allocate resources for performing the assessment
2. Establish or update the process reference model to support the assessment if needed
3. Map local processes to the reference model as needed
4. Document the scope of the management system to be assessed
5. Define assessment purpose, constraints, stakeholders and participants
6. Plan the assessment, schedule, resources, project plan and deliverables
7. Identify required data and evidence
8. Identify data sources
9. Conduct documentation review
10. Conduct workshops and interviews
11. Develop ratings and analyze results
12. Develop recommendations

d). Produce Assessment Deliverables –

1. Assessment Plan: Scope of required information to be collected, assessment activities, resources, schedule, participants, deliverable description
2. Data Collection: Data collection strategy, correspondence, gathered evidence
3. Data Validation: Identification of any data deemed unreliable or that does not meet the criteria of objective evidence
4. Process Capability Evaluation: Evaluation includes the rating, analysis, and recommendations

e). Prioritize and Recommend Improvements

Service Owners and Managers evaluate Service Quality and develop improvement plans. Improvement plans that require funding are submitted to the appropriate planning or portfolio management function for funding prioritization, capital planning, and investment control. Typically, the set of improvement recommendations are collated and analyzed collectively. The analysis of alternatives considers schedule, costs, risks, performance, value, and potential for disruption.

Step Four – Act

Take corrective action to continually improve

The purpose of this phase is to take the appropriate action to continually improve the Service Quality approach, and process capability improvement methods. This includes activities related to expanding and improving service management and process measurement capabilities as follows:

1. Establishing targets for improvements
2. Ensuring that approved improvements are implemented
3. Measuring implemented improvements against the established targets; acting where targets were not realized
4. Revising the Service Quality Plans and Procedures as necessary
5. Conducting Service Quality reviews annually
6. Service Quality reviews are recommended to be performed at least annually and include at least the following:

a. Follow up from previous reviews
b. Process capability assessments, improvements and performance reviews
c. Service conformity and evaluation results
d. Corrective and preventive actions
e. Customer feedback and complaints
f. Service Quality metrics and reporting results
g. Approved quality objectives and quality objective plans

11.2 Service Performance

We have covered Service Quality Management in detail and need to address the Performance Management of the IT Services. IT Performance Management is the use of performance information through monitoring and measuring relevant IT performance metrics. The information obtained from metrics enhances the organization's ability to gauge performance results. Actual performance can be compared with expected outcomes defined in organization goals and objectives in quantitative and qualitative terms.

The organization develops action plans and projects that are designed to achieve the goals and objectives of the organization hence the majority of IT performance management activities are monitoring actual IT performance against the organization's strategic plans. The results of performance management analysis enable leadership to direct the establishment and implementation of corrective plans to adjust and ensure that the organization achieves pre-determined levels of performance.

The linkage between organizational goals and objectives and its IT strategy are achieved if performance management is successful. As most leaders live this every day, there is a need for constant alignment – and re-alignment – between the organization strategy and the IT strategy and objectives.

Measures of performance must be designed to accurately capture the execution of relevant, measurable objectives. Establishing a baseline to assess against is crucial to performance management. Without an effective performance management program, key leaders are faced with taking corrective actions after performance issues impact the organization. Performance management as a management tool is the alternative to damage control and crisis management, which impacts organizational plans, goals and objectives, customer satisfaction, productivity, expenditures, and confidence in the IT Service Provider.

> Every organization's approach to implementing performance management is based on their IT alignment with the goals and the vision of its business customers. The following are general principles every organization should consider when developing an approach to performance management.

1. Improvements achieved through performance management should align with other management improvements within the organization: Organizations that generally change their focus from cost to overall outcomes and an emphasis on "value" appreciate the need for performance-based results of the organization and apply the same principles into IT performance metrics. The organizational management approach includes developing:

Strategic plans and mission statement

Identification of strategic goals linked to the functional group responsible for achievement of the goals

Specific planned actions to achieve goals

Performance plans that outline IT initiatives linked to the organization goals

Performance reports that inform leadership of the health of IT initiatives and links to future improvements decisions

Implementing and Improving ITSM

2. Implementing a performance management program takes time: There should be a realization that performance management programs are long-term initiatives and must be approached as a radical management shift in the focus of the organization and a change to IT practices.

3. Performance Management requires sustained management commitment and collaboration at all levels within the organization: A performance management initiative can only be successful when all levels of the organization's management provide unwavering support to the initiative and

4. The organization needs to ensure available resources:

Sufficient management and technical analysts should be dedicated to performance management with appropriate training and skills

Acquisition and implementation of performance management technology tools to capture data for analysis and store the information for historical purposes

11.3 Assess, Measurements, and Metrics

IT Performance Management metrics, once established through analysis of the organization's requirements for IT, should be implemented with leadership as champions.

> **Performance metrics encourage effectiveness, efficiency, and internal control of ITSM processes. Through periodic evaluation, metrics should be reviewed, updated or deleted as the organizational ITSM ability increases.**

Assessments

Assessments come in many forms and are available in most frameworks. Within a process improvement context, process assessment provides a means of characterizing in terms of the capability of selected processes. Process assessments are also used to identify strengths, weaknesses, and measure the extent that current practices are achieving the outcomes and purpose of the process. The advantage of conducting process assessments is the analysis of current practices and the output that provides recommendations for improvement that roll into a process improvement plan. Another version of process assessments may be the CMMI© 1-5 scale, which measures the repeatability of the process.

Measurements and Metrics

Every organization, IT and business, produces measurements and metrics to better understand quality and performance. A Measure is usually a raw number while a Metric is a threshold for monitoring the health and performance of the process. Below is an example of each based on the Incident Management process.

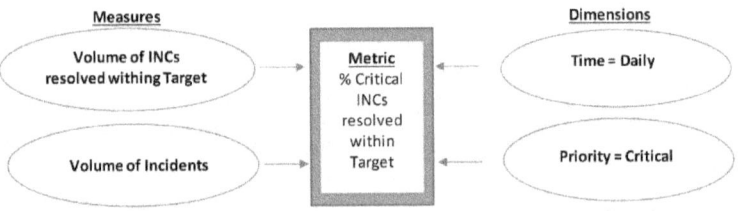

Metrics consist of one or more measures combined with a mathematical calculation and a standard presentation (format) for the output:

a. Metrics are associated with two dimensions, a time dimension and a functional categorization dimension
b. Metrics are used in the quantitative and periodic assessment of a process that is to be measured
c. Metrics should be associated with targets that are based on specific business objectives
d. Metrics are associated with procedures to determine the measures required and procedures for the interpretation of metrics results

Chapter 12 – Organizational Considerations

Implementing new policies, processes, and procedures within any IT organization affects the entire organization, including employees, customers, and stakeholders. To ignore the human side of change increases the risk of failure. Those responsible for ITSM must be aware of the potential impact on the people within an organization when implementing changes to policies, procedures, and processes. In short, those responsible for ITSM are organizational change agents who must build a bridge between people, processes, and technology.

Each year, there is a high percentage of ITSM programs, projects, and initiatives that fail due to unaddressed organizational culture. There is resistance to change within organizations. It is human nature. It is up to an organization's leadership to convey value to both employees and business customers for adoption. Employees gain value by using processes and standards for repeatable output. Business customers adopt when they feel included in the decisions made on their behalf. This is a tough concept for many within IT. Your IT organization is a Service Provider. These services should align with what the business customer wants to receive at a price they want to pay.

Organizational Change Management (OCM) provides a mechanism to address the human side of change. There are many OCM frameworks on the market and the best frameworks are those that are flexible enough to address general, specific, similar, and unique qualities of an organization's culture.

Chapter 13 – Risk Management

Risk includes all areas of organizations' environments, suppliers, business customers, and the individual people who make up the organization.

> Risk Management should be a consideration that includes people, process, technology, information, data, assets, infrastructure, governance, finances, and all other aspects that may impact the ability to meet the needs of

Best Practice supports the idea that risk should be incorporated in every phase of the Service Lifecycle and into how the individual processes are defined and improved.

> Risk Management should not focus on dealing with problems. Rather, it should focus on preventing them.

Looking at Risks through the Service Lifecycle, and identifying them, will help the organization manage risk effectively, reducing negative impact, uncertainty, and costs, and adding positive impact.

Glossary of Terms

*(Terms are taken directly from the ITIL© books).
Please see the ITIL© books for further explanation and context.*

Term	Definition
Access Management	Process responsible for allowing users to make use of IT services, assets, or data. Access Management helps protect the CIA (Confidentiality, Integrity, and Availability) of assets by ensuring that only the authorized users have access. This is called Identity and Access Management by many organizations. (Service Operation)
Accounting	Identifies the cost of delivering IT services, comparing these to budgeted costs, and managing the variance. (Service Strategy)
Active Monitoring	Monitoring of a Configuration Item (CI) or an IT service that uses automation to discover the current status. (Service Operation)
Activity	Set of actions designed to achieve a particular result. Activities are usually designed into processes and plans and are documented in the procedures.
Agreed Service	A synonym for service hours and is

Time (AST)	commonly used in calculating availability.
Alert	Warning that a threshold has been reached, something has changed, or a failure has occurred. Alerts are usually generated by a tool and are a part of the Event Management process. (Service Operation)
Asset	Any Resource or Capability. Assets contribute to the delivery of a service and can be: Management, Organization, Process, Knowledge, People, Information, Applications, Infrastructure, or Financial Capital. (Service Strategy)
Asset Management	Process responsible for tracking and reporting the value and ownership of financial assets throughout their lifecycle. It is part of Service Asset and Configuration Management (SACM) process. (Service Transition)
Attribute	A piece of information about a Configuration Item (CI). Examples include location, version number, and cost. Attributes of a CI are included with the CI in the Configuration Management Database (CMDB). (Service Transition)
Availability	Ability of a Configuration Item (CI) or IT service to perform its agreed-upon function

	as and when required. Availability is determined by reliability, maintainability, serviceability, performance, and security. Availability is usually calculated as a percentage. (Service Design)
Availability Management	Process responsible for defining, analyzing, planning, measuring, and improving all aspects of the availability of IT services. (Service Design)
Availability Plan	A plan to ensure that existing and future availability requirements for IT services can be provided in a cost-effective manner. (Service Design)
Baseline	Benchmark used as a reference point for measurement. This can be used for measuring service performance or for a Configuration Item.
Budget	A list of money an entity plans to receive and pay out over a given period of time.
Budgeting	Activity of predicting and controlling the spending of money, usually through a cyclical view (i.e., annual).
Build	The act of assembling multiple Configuration Items (CIs) to create part of an IT service. The term may also be used to refer to a Release that is authorized for

	deployment (i.e., server build). (Service Transition)
Business Case	Justification for a significant item or expenditure. It is the basis for decision-making. (Service Strategy)
Business Continuity Management	The Business Process responsible for managing risks that could severely impact the business. It involves the management of risk and is the basis for IT Service Continuity Management. (Service Design)
Business Continuity Plan	A plan defining the steps required to restore Business Processes following a disruption. (Service Design)
Business Customer	A recipient of a product or service.
Business Impact Analysis (BIA)	Activity in Business Continuity Management that identifies Vital Business Functions (VBFs) and their dependencies. BIAs are exercise and updated regularly. (Service Strategy)
Business Process	A process owned and carried out by the business.
Business Relationship Management	Process responsible for maintaining the relationship with the business customer. (Service Strategy)
Business	A role responsible for maintaining the

Relationship Manager	relationship with one or more business customers. (Service Strategy)
Capability	The ability of an organization, person, application, Configuration Item, or IT service to carry out an activity. These are intangible assets. (Service Strategy)
Capacity	Maximum throughput that a Configuration Item or IT service can deliver while still meeting agreed-upon Service Level Targets. (Service Design)
Capacity Management	Process responsible for ensuring that the capacity of IT services and the IT infrastructure can deliver the agreed-to Service Level Targets in a cost-effective and timely manner. (Service Design)
Capacity Plan	The Capacity Plan is used to manage the resources required to deliver IT services. (Service Design)
Change	The addition, modification, or removal of anything that could influence IT services. (Service Transition)
Change Advisory Board (CAB).	Group of people who aid the Change Manager in assessing, prioritizing, and scheduling Changes. (Service Transition)
Change Management	Process responsible for controlling the lifecycle of all Changes. The primary goal

	is to handle large volumes of Changes with minimal disruption to IT services. (Service Transition)
Change Model	Repeatable way of dealing with a particular category of Change, usually modeled after types of Changes. (Service Transition)
Change Request	Sometimes called a "Change Record," this is the record containing the details of a Change. (Service Transition)
Change Window	Regular, agreed-upon time when Changes and Releases may be implemented with minimal impact on IT services. These are usually documented in the SLAs. (Service Transition)
Component	General term that is used to mean a part of something more complex.
Component Capacity Management	Process responsible for understanding the capacity, utilization, and performance of Configuration Items (CIs). (Service Design)
Confidentiality	A security principle requiring the data to be accessed only by authorized people. (Service Design)
Configuration Baseline	Used as a basis for future build, Changes, and Releases. (Service Transition)
Configuration	Activity responsible for ensuring that

Control	changes to a Configuration Item (CI) are managed. (Service Transition)
Configuration Identification	Activity responsible for collecting information about Configuration items and their relationships, then loading into the Configuration Management Database (CMDB). (Service Transition)
Configuration Item	Any component that needs to be managed to deliver an IT service. (Service Transition)
Configuration Management	Process responsible for maintaining information about Configuration items (CIs) required to deliver an IT service, including their relationships. (Service Transition)
Configuration Management Database (CMDB)	A database used to store Configuration Records throughout their lifecycle. (Service Transition)
Configuration Management System (CMS)	A set of tools and databases (including CMDBs) used to manage an IT service provider's Configuration data. (Service Transition)
Configuration Record	A record containing the details of a Configuration Item (CI).
Continual Service Improvement	A stage in the Service Lifecycle responsible for managing improvements to IT Service Management processes and IT services.

(CSI)	(Continual Service Improvement)
Critical Success Factor (CSF)	Something that must happen if a project, process, plan or IT service is to be successful.
Definitive Media Library (DML)	One or more locations where definitive and approved versions of all software Configuration Items (CIs) are securely stored. (Service Transition)
Demand Management	Process whose activities attempt to understand and influence customer demand for services and the provision of Capacity to meet those demands. (Service Strategy)
Deming Cycle	Method for service improvement via four steps (Plan, Do, Check, Act). (Continual Service Improvement)
Dependency	Direct or indirect reliance by one process activity, IT service, or Configuration item on another.
Detection	A stage in the Incident Lifecycle allowing the results to be known to the Service Provider. (Service Operation)
Diagnosis	A stage in the Incident and Problem Lifecycles responsible for finding a workaround for an Incident and root cause for a Problem. (Service Operation)
Differential	Technique to influence the demand for IT

Charging	services in Demand Management. (Service Strategy)
Driver	Something that influences Strategy, Objectives, or Requirements.
Early Life Support	Support provided for a new or changed IT service for an agreed-upon period of time after it is Released. (Service Transition)
Effectiveness	A measure of whether the objectives of a process, service, or activity have been achieved. (Continual Service Improvement)
Efficiency	A measure of whether the right amount of resources have been used to deliver a process, service, or activity. (Continual Service Improvement)
Emergency Change	A Change that must be implemented as soon as possible. Usually, will be authorized by the Emergency Change Advisory Board (ECAB). (Service Transition)
Emergency Change Advisory Board	A subset of the Change Advisory Board who make decisions about Emergency Changes.
Event	The change of state that has significance for the management of a Configuration Item or IT service. The term may also be used to

	mean an alert or notification. (Service Operation)
Event Management	Process responsible for managing Events throughout their lifecycle. (Service Operation)
External Service Provider	An IT Service Provider that is part of an organization different from that of the customer. (Service Strategy)
Facilities Management	Function responsible for managing the physical environment (e.g., power, heating, cooling), physical access, and even environmental monitoring. (Service Operation)
Failure	Loss of ability to operate to specification, or to deliver the required output. A failure usually results in an Incident. (Service Operation)
Financial Management	Process responsible for managing an IT Service Provider's Budgeting, Accounting, and Charging Requirements. (Service Strategy)
Fit for Purpose	A term synonymous with "Utility." (Service Strategy)
Fit for Use	A term synonymous with "Warranty." (Service Strategy)
Follow the Sun	Methodology for using Service Desks and

	Support Groups around the world to provide a seamless, 24 x 7 Service. (Service Operation)
Function	A team or group of people, and the tools they use to carry out one or more processes or activities.
Functional Escalation	Transfer of an Incident, Problem, or Change to a technical team with a higher level of expertise. (Service Operation)
Hierarchic Escalation	Informing or involving more senior levels of leadership to assist in the escalation. (Service Operation)
Impact	Measure of the effect an Incident, Problem, or Change will have on Business Processes. (Service Operation, Service Transition)
Incident	Unplanned or threatened interruption to an IT service or a reduction in service quality of an IT service. (Service Operation)
Incident Management	Process responsible for managing the lifecycle of all Incidents. The goal is to resolve Incidents as quickly as possible. (Service Operation)
Incident Record	Record containing the details of a single Incident. (Service Operation)
Information Security	Process that ensures Confidentiality, Integrity, and Availability of an

Management	organization's assets, information, data, and IT services. (Service Design)
Information Security Policy	Policy that governs the organization's approach to information security and is used by Access Management. (Service Design)
Infrastructure Service	An IT service indirectly used by the business but is required by the IT Service Provider to provide IT services.
Integrity	Security principle that ensures that the data and Configuration items are modified only by authorized personnel and activities. (Service Design)
Internal Service Provider	Service Provider that is part of the same organization as their customer. (Service Strategy)
IT Operations	Activities carried out by the IT Operations Control function, including monitoring and control of IT services and IT infrastructure. (Service Operation)
IT Operations Management	Function within an IT Service Provider that performs the daily activities needed to manage IT services and supporting IT infrastructure. (Service Operation)
IT Service	Service provided to one or more customers by an IT Service Provider.
IT Service	Process responsible for managing Risks that

Continuity Management	could impact IT services. (Service Design)
IT Service Continuity Plan	A Plan defining the steps required to recover one or more IT services and should be part of the Business Continuity Plan for the organization. (Service Design)
IT Service Management	Management of quality IT services offered to the business customer. It is made up of people, processes, and technology.
IT Service Provider	Service Provider who provides IT services to internal or external customers. (Service Strategy)
Key Performance Indicator (KPI)	Metrics used to manage a process, IT service, or activity. (Continual Service Improvement)
Knowledge Management	Process responsible for gathering, analyzing, storing, and sharing knowledge and information within an organization. (Service Transition)
Known Error	A Problem that has a documented root cause and a Workaround. (Service Operation)
Known Error Database (KEDB)	Database containing all Known Error records. (Service Design)
Lifecycle	The various stages of the life of an IT

	service, Configuration item, Incident, Problem, Change, etc.
Maintainability	A measure of how quickly and Effectively a Configuration Item or IT service can be restored to normal working order after a Failure. (Service Design)
Major Incident	The highest category of Impact for an Incident. (Service Operation)
Management of Risk	Activities required to identify and control the exposure to risk that may impact an organization's business objectives.
Mean Time Between Failures (MTBF)	Metric for measuring and reporting reliability. Average of time after an IT service or Configuration Item starts working until it next fails. (Service Operation)
Mean Time Between Service Interruptions (MTBSI)	Metric used for measuring the average time between Incidents disrupting an IT service. (Service Design)
Mean Time to Repair (MTTR)	Average time taken to repair a Configuration Item or IT service after a failure. (Service Operation)
Mean Time to Restore Services	Average time taken to restore a Configuration Item or IT service after a failure. (Service Operation)

(MTRS)	
Metric	Something that is measured and reported to help manage a process, IT service, or activity.
Passive Monitoring	Monitoring of a Configuration Item (CI), IT service, or a process that relies on an alert or notification to discover the current results. (Service Operation)
Pattern of Business Activity (PBA)	A workload profile of one or more business activities used by Demand Management to understand and influence demand for IT services. (Service Strategy)
Post-Implementation Review (PIR)	A review that takes place after a Change or project has been implemented to identify lessons learned.
Priority	A Category used to identify the relative importance of an Incident, Problem, or Change. (Service Operation, Service Transition)
Proactive Monitoring	Monitoring that looks for patterns of Events to predict possible future Incidents. (Service Operation)
Proactive Problem Management	Part of the Problem Management process that analyzes Incident Records and other data to identify Problems. (Service Operation)

Problem	A cause of one or more Incidents. (Service Operation)
Problem Management	Process responsible for managing the lifecycle of all Problems, looking for ways to prevent Incidents from happening and minimizing their impact if unable to prevent. (Service Operation)
Problem Record	A record containing all the details of a Problem. (Service Operation)
Procedure	Document containing the steps that specify how to achieve an activity.
Process	Structured set of activities designed to accomplish a specific objective.
Process Manager	Role responsible for the operational management of a process.
Process Owner	Role accountable for the process being carried out as documented.
RACI	Model used to help define roles and responsibilities (R=Responsible; A=Accountable; C=Consulted; I=Informed). (Continual Service Improvement, Service Design)
Reactive Monitoring	Monitoring that takes place in response to an Event. (Service Operation)
Release	Collection of hardware, software, documentation, processes, or other

	Components required to implement one or more approved Changes to IT services. (Service Transition)
Release and Deployment Management (RDM)	Process responsible for both Release Management and Deployment. (Service Transition)
Release Management	Process responsible for planning, scheduling, and controlling the movement of Releases to Test and Live Environments. Part of the Release and Deployment Management process. (Service Transition)
Release Unit	Components of an IT service that are normally released together. (Service Transition)
Reliability	A measure of how long a Configuration Item or IT service can perform its agreed-upon function without interruption. (Service Design, Continual Service Improvement)
Request for Change (RFC)	A formal proposal for a Change to be made. (Service Transition)
Request Fulfillment	Process responsible for managing all Service Requests through their lifecycle. (Service Operation)
Restore	Acting to return the IT service or Configuration Item to users after an Incident

	(Service Operation)
Retire	Permanent removal of an IT service or Configuration Item from the Live environment (Service Transition)
Rights	Entitlements or permissions granted to a User or Role. (Service Operation)
Risk	A possible Event that could cause harm or loss or affect the business customer.
Risk Management	Process responsible for identifying, assessing, and controlling Risks, usually using a Risk Assessment.
Role	Set of responsibilities, activities, and authorities granted to a person or group.
Root Cause	The underlying or original cause of an Incident or Problem. Root Cause Analysis (RCA) will be performed to identify the root cause of an Incident or Problem. (Service Operation)
Service	A means of delivering value to customers by facilitating outcomes customers want to achieve without the ownership of specific costs and risks.
Service Asset	Any Capability or Resource
Service Asset and Configuration	Process responsible for both Configuration Management and Asset Management. (Service Transition)

Management (SACM)	
Service Capacity Management	Activity responsible for understanding the performance and Capacity of IT services. (Service Design, Continual Service Improvement)
Service Catalog	A database or structured document with information about all live IT services, including those available for deployment. (Service Design)
Service Design	A stage in the Service Lifecycle of an IT service, including processes and functions. (Service Design)
Service Design Package (SDP)	Document(s) defining all aspects of an IT service and its requirements through each stage of its lifecycle. (Service Design)
Service Desk	Function acting as the single-point-of-contact between the service provider and the users. The Service Desk owns all Incidents and Service Requests through the lifecycle. (Service Operation)
Service Improvement Plan (SIP)	Formal plan to implement improvements to a process or IT service. (Continual Service Improvement)
Service Knowledge	A set of tools and databases that are used to manage knowledge and information. The

Management System (SKMS)	SKMS includes the CMS and other tools and databases. (Service Transition)
Service Level	Measured and reported achievement against one or more Service Level Targets.
Service Level Agreement (SLA)	An agreement between an IT Service Provider and a customer. (Service Design, Continual Service Improvement)
Service Level Management (SLM)	Process responsible for negotiating Service Level Agreements and ensuring that these are met. (Service Design, Continual Service Improvement)
Service Level Requirement (SLR)	Customer requirement for an aspect of an IT service (usually based off of business objectives). (Service Design, Continual Service Improvement)
Service Level Target	A commitment that is documented in a Service Level Agreement. (Service Design, Continual Service Improvement)
Service Management Lifecycle	An approach to IT Service Management that emphasizes the importance of coordination and control across the various functions, processes, and systems necessary to manage the full lifecycle of IT services.
Service Manager	A manager who is responsible for managing the end-to-end lifecycle of one or more IT

	services.
Service Operation	A stage in the lifecycle of an IT service that includes processes and functions. (Service Operation)
Service Owner	Role that is accountable for the delivery of a specific IT service. (Continual Service Improvement)
Service Pipeline	A database or structured document listing the IT services that are under consideration or development, but not available for consumption. (Service Strategy)
Service Portfolio	The complete set of IT services that are managed by a Service Provider, including those under consideration or in development, live services, and retired services. (Service Strategy)
Service Portfolio Management	Process responsible for managing the Service Portfolio. (Service Strategy)
Service Request	A request from a user for information, advice, for a Standard Change, or for access to an IT service. (Service Operation)
Service Strategy	A stage in the Service Lifecycle of an IT service, including processes and functions. (Service Strategy)
Service	A stage in the Service Lifecycle of an IT

Transition	service, including processes and functions. (Service Transition)
Service Utility	Functionality of an IT service from the customer's perspective. (Service Strategy)
Service Validation and Testing (SV&T)	Process responsible for the validation and testing of a new or changed IT service. (Service Transition)
Service Warranty	Assurance that an IT service will meet agreed-on requirements (Capacity, Availability, Security, Continuity). (Service Strategy)
Serviceability	The ability of a third-party supplier to meet the terms of their contract. (Service Design, Continual Service Improvement)
SLAM Chart	Service Level Agreement Monitoring Chart used to monitor and report achievements against Service Level Targets. (Continual Service Improvement)
Standard Change	A pre-approved Change that is low risk, relatively common, and follows a procedure or work instruction. (Service Transition)
Supplier Management	Process responsible for ensuring that all contracts with suppliers support the needs of the business and all suppliers meet their contractual commitments. (Service Design)

Threshold	The value of a metric that should cause an alert to be generated or management action taken.
Transition Planning and Support (TP&S)	Process responsible for planning all Service Transition processes and coordinating the resources they require.
Types of Service Providers	Type 1: Internal Service Provider embedded in a business unit Type 2: Internal Service Provider that provides shared IT services to more than one business unit. Type 3: Service Provider that provides IT services to external customers (Service Strategy)
Underpinning Contract	Contract between an IT Service Provider and a third party underpinning an SLA the IT Service Provider has with the business customer.
Urgency	Measure of how long it will be until an Incident, Problem, or Change has a significant Impact on the business. (Service Operation, Service Design, Service Transition)
Usability	Ease with which an application, product, or IT service can be used. (Service Design)

Vital Business Function (VBF)	Function of a Business Process that is critical to the success of the Business. (Service Design)
Vulnerability	A weakness that could be exploited by a threat.
Work Instruction	A document containing detailed instructions that specify exactly what steps to follow to carry out an activity.
Workaround	Reducing or eliminating the Impact of an Incident or Problem for which a full resolution is not yet available. (Service Operation)

www.ingramcontent.com/pod-product-compliance
Lightning Source LLC
Chambersburg PA
CBHW071041240526
45471CB00014B/154